It Isn't Just Business, It's Personal

HOW PAETEC THRIVED
WHEN ALL THE BIG TELECOMS COULDN'T

by Arunas A. Chesonis and David Dorsey

RIT CARY
GRAPHIC ARTS
PRESS

ROCHESTER, NEW YORK

It Isn't Just Business, It's Personal:
How PAETEC Thrived When All the Big Telecoms Couldn't
Arunas A. Chesonis and David Dorsey

Published and distributed by
RIT Cary Graphic Arts Press
90 Lomb Memorial Drive
Rochester, New York 14623-5604
http://library.rit.edu/cary/carypress

The views expressed in this book are those of the authors
and do not necessarily reflect those of Rochester Institute of Technology.

Book design by Marnie Soom / RIT Cary Graphic Arts Press
Cover design by Suthida Sakulsurarat / PAETEC Communications, Inc.
Printed in the United States.

ISBN-10 1-933360-18-6   ISBN-10 1-933360-21-6 (cloth)
ISBN-13 978-1-933360-18-8   ISBN 13 978-1-933360-21-8 (cloth)

Library of Congress Cataloging-in-Publication Data

Chesonis, Arunas A.
      It isn't just business, it's personal : how PAETEC thrived when all the big
   telecoms couldn't / by Arunas A. Chesonis and David Dorsey.
         p. cm. — (First person corporate profile series)
      Includes index.
      ISBN-13: 978-1-933360-18-8 (pbk.)
      ISBN-10: 1-933360-18-6 (pbk.)
      ISBN-13: 978-1-933360-21-8 (hardcover)
      ISBN-10: 1-933360-21-6 (hardcover)
      1. Telecommunication—United States—Management. 2. Management—
   United States—Employee participation. 3. PAETEC Communications.
   I. Dorsey, David, 1952– II. Title.
   HE7785.C44 2006
   384.054'73--dc22

                                                      2006035114

Contents

Foreword

WHAT IS TECHNOLOGY? Answer: Tools, processes, methods, techniques, and systems developed by people for people.

What is profit? Answer: The left over after all costs have been covered by revenue received by people, from people, and for people.

What is an organization? Answer: A collection of physical assets and technology, and a reporting and production structure designed by people to serve people.

Who are we? Answer: We are people, that is, human beings who have ambitions, passion, and feelings which reflect life, substance, and values through interactions and connections with other people.

Exploring these answers is what this book is really about. Of course, the book talks about how an idea spawned a company, drawing on a keen entrepreneurial spirit, and how the "company" figured out how to compete in a dangerous, ever changing, and fast moving world environment. But what is the underlying formula for the company's success?

The formula centers around people. People have many attributes in common: for example, the need to belong; the joy of giving; the adrenalin burst when team effort leads to a win; the knowledge that no one can do everything – even anything – alone; the desire to receive respect from others and the obligation to offer respect to others.

At the same time, each individual is unique. Some individuals try so hard to blend in with the "organizational culture" that they lose their individuality and the creativity and ego necessary for success. Then there are individuals who try to go it alone, without listening or trying to understand what others might be saying or feeling.

Finally, we have people who think outside of the box, sometimes exhibiting a "crazy" personality or "quirk," in this way emphasizing and maintaining their individuality and uniqueness. Often they do what they do to make others feel at ease or to introduce humor into a situation. However, when they need to work together and care for and support one another, they are able to do that. That balance between being part of the human race and serving and working with people, while at the same time maintaining loyalty to their own individuality so that originality and uniqueness can come to the fore, is a formula for success. It is the formula for PAETEC's success.

The most complete and authentic reflection of this formula takes the form of an individual named Arunas Chesonis. He is smart. He is quick. He works tirelessly. I came to know Arunas first as an entrepreneur and community leader, and then as a friend who generously agreed to become a Trustee at the Rochester Institute of Technology.

Arunas also puts family first. He extends family to beyond his parents, wife, and children to those loyal employees of PAETEC and its vendors and customers. He takes time to relax and have fun with his immediate family and his extended family. He believes and trusts his extended family and expects that, utilizing their own individuality with a sense of shared purpose, they will do what is best for themselves and PAETEC. Of course, what is best for each individual and for PAETEC as a whole is the same thing.

Arunas represents the ideal contradiction of an inspirational entrepreneur who succeeds by being the outstanding servant leader. He is the model that is reflected in employees up and down the PAETEC organization.

This book is about that model. The model describes the four fundamental principles of:

- A Caring Culture
- Open Communication
- Unmatched Service and Support
- Personalized Solutions

There are lessons here for every organization – profit and not-for-profit. After all, we are all people and this book is about how people can succeed.

Dr. Albert J. Simone
President
Rochester Institute of Technology

Introduction

THOUGH YOU'VE PROBABLY NEVER HEARD OF PAETEC Communications, Inc., by most measures we're the most successful telecom of the past decade. As a full-service telecommunications company based just outside Rochester, New York, we come out on top in almost any comparison against other little companies you don't hear much about, and the big guys you may wish you'd heard a little less about: WorldCom, Global Crossing, and the rest. Like all the others, we offer telephone, data, and Internet services. That's most of our business, very commonplace stuff. And, from the standpoint of raw numbers, our success at this point is still pretty modest. Our revenues are only half a billion, our employees number less than 1,400, and our scope is, in many respects, purposely narrow. We serve less than one percent of our market.

This is another way of saying there are many things we've chosen *not* to do. We don't originate wireless service, though we often arrange and manage it for our customers. We don't serve residential homes, and we don't try to satisfy all the needs of huge organizations. We aim for the underserved enterprises, mid-sized outfits, as well as small segments of business at big names you already know: JP Morgan Chase, Cisco, Citigroup, Princeton, Duke, Hyatt and thousands of others. All of this cuts out large slices of the telecom pie. Which is the whole point. We stay focused on the people we serve.

Focusing on *people*. Making it *personal*. Our mission statement is to be the most employee- and customer-oriented telecom in the industry. We've turned this into an art, and it's precisely how we've managed to grow more rapidly than superstars like MCI, before Verizon merged with it, and even more dramatically than famous startups outside telecom, like Dell and Amazon. Qualcomm wouldn't qualify to carry our lunch box if you compare the first six years of its growth to ours. Let me be specific.

In 2003, PAETEC ranked second in the Deloitte Technology Fast 500, a list of the fastest-growing technology companies—not just telecom, but technology—in North America, based on average percentage revenue growth over five years. From 1998 through 2002, PAETEC grew roughly 192,000 percent, and even without the first two years of explosive expansion, our growth rate is still 250 percent. In 1998, when it was founded, PAETEC's payroll included eight founders and a few dozen additional people. Our dramatic growth is the product, in part, of a corporate culture that remained steady during the years of our greatest growth. Everything at PAETEC revolves around respect for people—the employee and the customer. The word customer may be a little more prominent in the mission statement, but, in reality, PAETEC puts employees first, and then watches them voluntarily put customers before themselves.

What we do isn't just business, it's personal. It's about the people we hire, the people we serve, the people we help in ways that extend beyond the daily grind of providing products and services. Business isn't a repetitive mechanical process for us; it's about creative human relationships. Anyone who wants to make a profit must learn how to make work meaningful to people, and the best way to do that is to put them first. We've done this. Though it gets tougher to maintain the culture and spirit as we get bigger, working at PAETEC still means something to its workers beyond the individual bottom line of salary and bonuses.

In an era when the CEOs of major American corporations, including telecommunications giants, were being indicted and convicted for fraudulent practices that enriched top executives at the expense of shareholders and employees, PAETEC was consistently attempting to make an honest profit by doing the right thing in order to build loyalty among employees, customers, and vendors. Three times we've had a chance to go public, but every time we've turned it down—because we didn't believe it was the right thing to do for our people and our shareholders. In each instance, the stock would have begun trading at a value too low for them to make what they deserved on their stock options. Those of us at the top would have made out just fine, but the rest may have felt betrayed. Therefore, we're still a private company.

It Isn't Just Business, It's Personal will describe a matrix of practices that have become the engine of our success: caring culture, open communication, unmatched service and support, and personalized solutions. I will describe the principles behind our success and tell how our people have put those principles into action. Much of this can be done at any company. Some of it can't. Some things you just can't duplicate. For example, the people at PAETEC are an unusual bunch. On Easter, I often dress as a large rabbit for the children of our employees who hunt eggs around the office. Need I say more? When various events give us an opportunity to bring in the children of our people, we'll hook up a Playstation 2 and let them compete at video games using the giant monitor in our Network Operations Center. We have one technician who, during down times, used to bring motorcycle parts to his desk and work on them, while wearing the ruffled shirt from a tuxedo. Our head of marketing, Jack Baron, is a guitarist and lead singer in his own rock band, Sometimes Three, and has a complete sound studio in his home. Dan Reinbold, who runs our employee training and is a karate black belt, often emphasizes a point during one of his presentations by splitting wood or cinder blocks with the blade of his bare hand.

At our best, we're a close-knit family. This is reflected in the unusual way we came up with the name for our company. Before we started out, we considered naming this place Polaris. But that domain name had already been taken, so we kept mulling over different ideas. One evening, we opened a bottle of wine to share with two friends who were among the first people to join us: Molly Korndoerfer and Kathy Hill. They suggested we name the company after my family. They'd come up with PAETEC, but we were missing an "e." My wife, Pam, happened to be pregnant at that point, and so I called her and asked, "Can our next child's name begin with an e?" She said, "How about Emma, if it's a girl." That's how Emma got her name, and how PAETEC got *its* name: Pam-Adam-Erik-Tessa-Emma-Chesonis.

In stressful times when it may not seem as much of a family as it once did, we are, at the very least, a collection of unusual individuals who have succeeded because we value our individuality and the human bond that has held us together from the beginning. The key is to encourage this bond to extend outward to include customers. Our four principles—caring culture, open communication, unmatched service, and personalized solutions—have enabled us to create and strengthen that bond over the years.

If you put in a day, a week, a month at PAETEC, you begin to realize what people are actually trying to keep alive here. It isn't just about money. It's a personal connection with other workers and with the company's mission, at a level of emotional commitment they haven't found anywhere else. You can list all the quantifiable elements of the company's success, but this intangible spirit, the aura of something special and unique unfolding, is what matters most here. The more we grow, the tougher it gets to maintain this feel, this sense that we are special, that we're a family, and that we care about one another the way only people in an entrepreneurial venture care. But despite the growing pains, the hairline fractures in our spirit that come as a natural consequence of success, there is still, underneath it all, an unusual kind of goodwill that flows up and down the organizational chart, and horizontally as well. You have the sense that each employee is doing everything possible, going to any length, not to spoil this goodwill. As one of my top people once put it, we do everything "Not to soil the PAETEC name." And that's why our customers stick with us.

One thing hasn't changed since Day One: we continue to value people before quick profit. Our mission is to be the most employee- and customer-oriented telecom in the industry. It's the most basic axiom of our success, a genuine example of how the people who lead the company think on behalf of everyone who works for PAETEC and not just for themselves. It's the fundamental principle behind all the others in this book, and the one I want to talk about before all the rest: Put People First.

Arunas A. Chesonis
Rochester, New York

PART ONE: CARING CULTURE

CHAPTER I

Put People First

IT'S THE PEOPLE, STUPID. Live by that slogan, and you can get elected to office, run a corner grocery, throw a great dinner party—and even grow a $500 million telecom with a little help from your friends. It's how we grew PAETEC. It's the people: employees, customers, suppliers, and everyone else in the communities we serve. Having a caring culture means your business begins and ends with people, not a logo, or a business plan, or a domain name. As a result of this fundamental principle, the first hundred people signed up before we even clearly knew how we were going to do business. They knew their worth would be recognized and that they would be joining a family of co-workers.

I became committed to the notion of putting people first when I worked for ACC Telecom in Rochester, New York, before it was sold to AT&T. In 1998, I could see the opportunity to leave ACC and start a new company, with a clean slate—a chance to avoid all the mistakes I'd seen a telecom make in the past and do things the right way, from the first day. The most important lesson I'd learned was that success in business begins and ends with people: the ones you hire, the ones you partner with, and the ones you serve as your customers.

It isn't simply the right thing to do. It's the *only* way to do business now. Everything, telecom no less than anything else, is becoming a commodity. As Thomas Friedman points out in *The World is Flat,* the playing field has become level for nearly everyone: any traditional advantage a company or a nation or an economy has had over others in the past, is disappearing. Knowledge is freely available. So is talent. So is nearly every resource that used to give one enterprise an edge. What remains is the quality of the human relationships that hold an enterprise together. An enterprise needs to put

3

people first, so that its employees will do the same thing in the way they relate to customers. It's management's responsibility to recognize that its greatest competitive advantage now is the way its people work for one another and for the market they serve. In the flat world, if business isn't intensely personal, it won't succeed.

HIRE THE RIGHT PEOPLE

PAETEC looks for *character*. Work ethic. Team spirit. A willingness to admit mistakes and strive to improve. That's the core competitive advantage: people who *care*, who value human relationships above everything else, relationships in the workplace, with customers, in the community and at home. So the first step in putting people first is to hire people *who put people first*, to hire for character—for a level of caring consistent with our culture—and not just experience or intelligence or the usual indicators of performance.

This means you'll have a team of people who will stay true to their character when they might be most tempted to depart from it: during those years when everything and anything seems to make money—if those years ever return—as well as during the downturns when almost nothing does. Without hiring the right people, it doesn't make sense to treat them as the company's most important asset. It all starts with the most rigorous hiring program possible, the first step in asking people to behave as if they are participating in an entirely new way to do business, something delicate and easy to spoil. PAETEC often has dozens of open positions in sales because it can't find the right people to fill them.

It isn't for lack of applicants. It's just that PAETEC is fanatically picky about who it hires. We have a long list of people who want to work for PAETEC in our Rochester headquarters. Though few people know who we are in most of the markets we service, in Rochester we've become a household word for many people, partly because of the extensive ways we've been involved in promoting the community and partly because we're known as a great place to work. People line up to work for us.

The few who get hired are picked because PAETEC hires first for common sense, a common touch. When it checks references, the company wants to know: does this person operate with common sense? Is he or she a quick study? What about attitude and interpersonal skills? Is he or she willing to put an ego on hold and serve others without seeing an immediate personal payoff? Track record isn't the most important thing because the person is coming from the ordinary world into something entirely different; how the person performed in the old world isn't a true indication of what he or she can do in the PAETEC world. Is this the kind of person you invite over to your house for dinner on Friday? Will this person fit into a community of, essentially, best friends? It's important because PAETEC hires people it plans on keeping for a long time and promoting internally.

PAETEC puts good people together and then watches as good processes flow into place. Work ethic, integrity, intelligence, and interpersonal skills: those are qualities that are either there or not there in a recent college grad and a twenty-year veteran alike. Oddly enough, good times can be the toughest test of character. That's when people are naturally eager to bask in rewards for their effort in ways that would erode the character of the company. I've got people who want the extra perks now that we're successful, and I still say no.

Keith Wilson, our CFO, says an employee's character has to be solid *before* we hire: "At the end of the day, there are ethical challenges all the time. Of course there will always be numerous opportunities to take advantage of things that would be borderline gray or unethical. If you violate anyone's trust, you ruin your career here. That applies especially in hiring. You hire good people to do the right thing. You can't hire people who say, 'I'm an ethical person now that I'm at PAETEC.' You have to be a good, ethical person to begin with."

I had a distinct advantage in hiring right from the start. For many years, I had worked with almost all of the first group of people we brought on board. I knew them, personally and professionally. And they all knew, and liked, one another. We were already a family before we ever rented space to work in. That was a huge leg up on the competition, and not something just any start-up could replicate.

The way we hired our first group of people built excitement and motivation, and increased the sense of team spirit. Before we began the interviews, we wrote down the names of seventy-five people we wanted to bring on a whiteboard in red marker. As individuals accepted our offers, we changed the color of their names on the whiteboard to green. When new people came for an interview, they saw, in green, the names of people who were already on board, and it created for many of the new interviewees a sense of wanting to be included, of wanting to belong to something new and exciting, of not wanting to be left behind. It built enthusiasm and made it easier to hire, as the list got bigger and bigger. When the names of the key people we wanted to hire in engineering and provisioning were in green, it gave all the rest of us a huge sense of confidence in our potential for success, much of which hinged on having some of the best technical people in the industry. The people we wanted to bring on board also knew those names. Most of them had worked together before.

It didn't hurt that most of them were experts in their fields: finance, engineering, sales, IT, and customer service. We never could have succeeded without the street smarts they brought with them, along with a deep knowledge of both the technology and the way telecom works. So, obviously, you have to start with a core group of people who can do their jobs in their sleep, and then bring in people who may be less experienced, but have the character necessary to build a caring culture. Dick Ottalagana and I sat down one day, right after we left ACC, and sketched out the notion for PAETEC on the back of a napkin—and we knew, before either of us started talking, that we would succeed, even though we didn't really know how the business would be structured. We knew that whatever we did, it would work, because we could invite three or four dozen first-string people to be a part of something new—some of the most talented people in the industry—and they would follow us. We already had that kind of trust and loyalty. Why?

I'll let Bob Moore, our head of IT, describe what it was like working with me at ACC. Bob is young, animated, highly intelligent, stubborn, and a former Eagle Scout. He's a guy willing to go with his instincts and admit his mistakes, and he's one of the key people who have made our success possible. Here is what he had to say to my co-author about working with me. I'm quoting him because what he says applies to everyone who works at PAETEC: the qualities he attributes to me are what we look for in everyone.

I'd gone to school here, and my first choice was to stay in Rochester. The company I was working for, Ambix, doing software, got into two projects with ACC. Arunas was president of ACC, very young, and he came down, and we worked on a project in the United Kingdom. One of his strengths was that he started another division outside the United States. He'd started the UK division. They had problems. They asked us to solve the problems, and we did. All through that effort, it was like working with a partner, someone at the same level—I was working *with* him, not *for* him—even though his company had hired me to work for him, in reality.

That was my first real exposure to Arunas. What I remember about those meetings was, he would come down, and ask us, "Hey how did it go?" I'm, like, twenty something, typical developer, jeans and T-shirt. We had one guy, who's here at PAETEC now, who wore fatigues every day. We had a couple of those meetings, and we were basically really honest. He appreciated honesty. Arunas sliced through the bull. Arunas asked us a tough question. I'm twenty something, remember. So he asks, "What would you do to fix this problem?" I'd say, "I'd do bang bang bang." The president of my company is over in the corner sitting like this, "Oh my God," wringing his brow. You could see Arunas was really, genuinely interested in whether or not we could help his business, even if what I said didn't reflect well on what ACC had been doing in the UK. It didn't matter. He *listened*.

Second, we were doing this project for Rochester Institute of Technology. ACC was doing a project but didn't have software to meet the demands, and so Arunas worked with me on that. We jointly went over to RIT and did some things. It was the same kind of relationship. The way you would work with an equal, not a superior. I was sitting in the car driving over with the president of ACC, Arunas Chesonis, and we were strategizing together, like old buddies, and I was this twenty-something snotnosed code jock. He's like, "Hey, why don't you ride over with me." You could see his mind working. He was, and is, very down to earth in meetings. Cutting right through it, "What's it going to take to get you happy? Bob, can we do that?" "Yep." "OK, that's what we're committed to."

I let Bob lead the effort, and he knew it. The point here is respect for people— the caring culture—along with the expectation of leadership from everyone. As I'll talk about more extensively later in this book, we're a company full of leaders, not followers, and I, as CEO, am willing to follow the lead of anyone who knows better than I do what should happen. It's the way we manage. Make everyone understand that he or she is just as crucial to the success of the organization as anyone higher up in the organizational chart. When you create that kind of culture, you can draw the best talent, because it's the way everyone wants to work. And when you bring together a core group of stars, the rest will follow. When we brought on board our key people, our founders, the rest knew it was safe to sign up.

Donna Wenk, senior vice president, put it this way:

I didn't jump on board immediately. I was a hold-out for a little while, and Arunas still kids me about that. I came here ultimately because of Dick Padulo, our VP in charge of operations and engineering. When you have a Dick Padulo in charge, you know it's going to work. Dick Padulo knows how to build a switch site and how to operate a Lucent 5E, our platform switch, better than anyone in the industry. All the 5E engineers I've known came here because of him. I knew he'd get the switches built. Dick will make it happen.

The point: bring in the key players first and they'll draw in a team of people who will lead from all levels in the company. The people are the whole ball-game. Treat them in a way that reflects the truth of that.

A CARING CULTURE ISN'T SOFT STUFF

Caring culture means never losing sight of the fact that your people matter more than the numbers. The people come first, and the numbers follow. You have to set everything up so that people know how much they matter to you, and that the company lives or dies by the passion for quality they bring to their work. That you care about them and their work in a way that isn't just about the numbers. The results need to be at the top of everyone's mind—whatever metric a given person needs to achieve—but only because the human relationships make it matter.

Not all CEOs speak this language. It doesn't mean they are bad CEOs. It just means that sometimes the numbers do have to come first, but this hasn't been the case at PAETEC during the years of our most rapid initial growth. Jeff Burke, our executive vice president, recalls how one founding employee reacted to Jack Welch when the famous CEO came to Rochester to speak:

> We're at the dinner where Jack Welch, the CEO of GE, had been flown into Rochester to speak. We're sitting at this table with Welch. John Budney, the VP of agent sales for PAETEC, John Morphy, the CFO of Paychex, and Steve McCluski from Bausch and Lomb are sitting at the table. Everyone was hanging on every word that Welch said. You would have thought he was the burning bush. We asked him what boards he participated on and he said, "I've never participated on any other boards. I'm a very focused individual. I'd take an operating role and end up running the company on any other board. I couldn't afford to do that."

> At one point, Welch was telling us how he took GE from a $3 billion company when he started to a $100 billion company. Budney looks over at him and says, with a smile "Oh. Only $3 billion." When I asked about how he'd implemented Six Sigma, he said, "Don't get caught up in Six Sigma. It's a tool. Use what's useful for your company. If you use it, do the light version." And that's what we ended up doing, as a matter of fact. But all along, Budney was squirming

through this whole thing. He was very impatient with the way we all treated him like an idol. Finally Budney turns to him and says, "Yeah, yeah, yeah. We can read all about Six Sigma in your book. We're paying you good money to tell us something useful. Tell us something we don't already know."

Welch sort of did a double-take and smiled, despite himself, and made the shape of a gun with his fingers and fires off a little shot of admiration in John's direction. By the end of the evening, they were sharing a drink together, like old friends. Welch enjoyed Budney's nerve, but he didn't really get Budney's point. John was trying to get him to talk about more than the numbers. A caring culture was not what GE needed at that point. He talked about the guy who ran Merck who said, "We have an obligation to make great drugs." He told us, "I had an obligation to make great numbers. That's what I did every day." It worked for GE, but in a place like PAETEC, if you always put numbers above people, people would stop caring, and customers would start going elsewhere for what you're selling. Because the caring is primarily what we're selling at PAETEC. And these days they can go *anywhere* for what you're selling, no matter what it is. You have to care about your people if you want your people to care about customers.

Often in a caring culture, the quality of the human bond gets expressed most effectively down in the details. This may sound a little odd, but we literally started from the ground up. Our caring culture is designed into the structure of our headquarters—the actual architectural style of the building. I hired my sister, Jolanda, to run human resources, and one of her additional jobs was to oversee the design of our offices. Here's what she has to say about those early duties:

> When Arunas tried to convince me to come on board he knew he wanted me to be involved in building the corporate headquarters. He gave me a couple charges and let me go. For internal people, we wanted the building to feel as if you were walking into your house coming through the front door. Second, for new customers and prospects, we wanted them to get a good sense of the company, to see we're not operating out of a shoebox. We did a lot of things to make it look expensive without putting that much money into it, using different paint colors, different carpet colors, choosing everything to carry the homey theme through. Arunas loves board games. So we themed our conference rooms after the Clue game. The Music Room, the Map Room, the Hunt Room, the Chess Room, the Sports Room, the Library. You feel as if you're doing business in someone's home.

We used wood finish and other touches in the NOC [Network Operations Center] to make it warmer. We designed the interior space so that outdoor light would shine through windows into the center of the building, so that even people in cubicles would have natural light. In most other offices, the managers and VPs have external offices with windows, and workers never see the light of day. We wanted to make sure we didn't do that. They are sitting in their workstations eight hours a day. We wanted to provide natural light and bring the outside in. We made sure there were no executive corner offices. We opened it up with glass walls inside the building so that light flows through as much as possible. And we have standard sizes in the offices for managers, directors, and VPs.

This sort of workspace was a huge change for the first few hundred people—and a huge investment in the well-being of our entire workforce for years to come, here in Rochester. By the time we established the business enough to build a facility like this, our people had learned to work in the most constrained of circumstances. We were all sharing space—low-rent, orange-crate-and-folding-table space, basically—during the first couple years. This was in the dot-com era when companies were going public and installing granite countertops in their rest rooms before they'd figured out how to make a profit. Some of us worked out of an old Kodak facility known as the IOT building, which stood for It's Only Temporary. There was no such thing as an office supply closet. We were all shooting from the hip, making leadership decisions at all levels, sleeping on bubble wrap in the back office if an emergency required a stretch of forty-eight consecutive hours. Nobody questioned the work load. We were running on empty, not making a penny that entire first year, and yet no one displayed anything but passion for the success of our start-up.

Why? The quality of the people. That spirit has lasted through all the years of our growth. As we became more organized and began to get our switches up and running, the scrappy spirit just got stronger. Budney referred to PAETEC people as "mudders," horses who ran with even more spirit and determination in the driving rain.

Wendy Showers, our director of account development in Irvine, California, recalls:

> I was hired as a manager of account development in Southern California. We didn't have furniture or customers or even a switch. We bought service from other telecoms and marked it up a little. I came on board to help bring the customers in. I can't think of anything I didn't do to win over a customer. We all just rolled up our sleeves and helped out wherever somebody else needed help.

I vacuumed the carpets. I emptied trash. We had a really green sales team at that time, a bunch of kids with a ton of enthusiasm, but not much experience. And PAETEC was, obviously, new to Southern California. It was hard to get people to sign up to sell for us. I've been in telecom for ten years, and I've worked for companies like WorldCom. I had five years of experience when I joined in 1999. I bought my own pencils. Everybody bought their own pencils: it's something we still talk about. If you wanted a pencil you went to Staples and bought one. I came up with how we manage customer files, and I helped manage the installations at customer sites. We had a calendar, before Outlook, so I was managing installs on the calendar. I did order processing functions. We would get involved, and I would be the one who scrubbed all the orders, not the sales manager. I'd go through it piece by piece making sure it was right. I was in sales! All of this, and I'm not even getting into what I did in sales.

In the same vein, Dick Padulo recalls, "I used some of my own personal frequent flyer miles to fly, to take business trips. There was no money coming in and we didn't have a lot of cash, so we did that, we all did that. We didn't splurge on anything."

We expect this kind of volunteerism, if you will, inside the organization, outside the job description. Everyone's job description here is pretty much the same in one respect: it's implied by the words *PAETEC employee*. You might have been hired to paint the fire hydrants, but if you want to stay at PAETEC, you'll never say, "Hey, my job is painting fire hydrants, I'm not responsible for putting out fires." Everybody is responsible for everything. Everyone is expected to keep the big picture in mind, and if you can contribute in some small way, no matter what department it might fall under, you offer your abilities. We don't need to tell people about this. If we've hired you, and you're the right person, you'll figure it out within a day or two: you can tell what's expected of you simply by walking in and putting in a day of work. This willingness to think and act outside the box of a job description will be happening all around a new hire—it's impossible to miss. Everybody will be doing it.

People have been willing to work this way because we treat employees with respect from the first day. We hire people and grant two weeks vacation immediately, period. For the first three years, we provided free lunches for all employees. As Molly Korndoerfer recalls: "In the beginning, it was five or six people splitting a Wegman's sub. 'Come on, let's eat, Arunas, lunch time. Let's cut up the sub.' When more people came on board, Arunas decided, 'We're still going to have lunch.' We discontinued it when we got up to a couple hundred people. By then we were serving hot meals. Nothing super fancy, but it was unheard of."

Making things personal goes much deeper at PAETEC though. It's in the spirit of the way we want everyone to manage those who work for them. Everyone who works as an executive or manager is expected to do everything possible to help his or her people succeed in their careers. A great example of this comes from a former NOC manager:

I had one employee who didn't have the technical background for the NOC. We liked his attitude, he was honest and forthcoming. But after three or four months, he was moving in circles and we didn't think he'd make it. I was in a meeting, and someone asked if we should pull the trigger. It was quiet for a second. I said, "I'm going to feel awful if we do. I don't think we've done everything. It's a tough job. We expect everybody to sit down and just go to it. Here we are talking about firing him." As soon as I said that, his fate became my project. I started to refocus on this one employee. I went and sat with him as soon as I got in the next day, and I spent the bulk of a day with him. The effect this kind of attention had on him was amazing. His confidence emerged. After two weeks, he got to the point where he was unbelievable. Before, he knew what he was doing, but he didn't have confidence and he just needed somebody to tell him, "Yes, you know what you're doing. Go ahead and do it." He crawled out of his shell. His attitude changed. It was almost embarrassing. For a month straight, every single day when he left the office he shook my hand. And we had been talking about putting this guy on unemployment! Later, at a team-building scavenger hunt, he was the king. Everybody was astonished. The confidence was overwhelming. Now, I can't imagine his not being a part of the team.

It would have been much easier for a manager with this person's title, workload and schedule to just give that young man his notice and thank him for his effort. But, instead, this manager knew his employee was trying his hardest, and so he felt it was his personal responsibility to this other human being to do whatever he could to help him succeed. This is a quality common to people we try to hire: it's a key element in our caring culture. Obviously, there are limits, but the limits at PAETEC allow for a lot more caring than at most companies.

Care and respect for our employees, putting them first, runs throughout all the principles of this book: open communication, unmatched service, and personalized solutions. Each of these other three fundamental principles requires the passion for quality that a caring culture creates. The caring culture is embedded in the way we compensate people, the way we encourage a balanced work life, the way we provide all of our employees with stock options, and the way we listen to even the most offbeat ideas on how to improve things. When an employee has a baby, we write notes of congratulations with a $500 savings bond included. When someone has a death in the family, we send our condolences. We write these things down instead of just picking up the phone. Something written endures—and it's there for everyone to read. As one of our top people puts it: "You have to give a damn. At most companies, employees really don't give a damn, because the company has made it ever so clear they don't give a damn about *them*. This isn't about money. It's a handwritten note from Arunas or from me to say, 'I'm sorry to hear about your loss,' when the person's mother dies. You may have to make a special effort to write a note for anyone who works for you, but it's a core competency, doing that."

It's also, maybe most importantly, implicit in how a personal emergency trumps all other duties at PAETEC. For example, when we were just starting up, one of our founders, under the stress involved in those first few months, developed severe back spasms and was urged to go for surgery to alleviate his back pain. He didn't want to. He wanted to recover without surgery. When he told me this, I told him to work from home for half a year. We held meetings in his home, in his dining room, where he would hold forth while stretched out on his back, on the floor. It was impractical for us, but none of us were suffering the kind of pain he was going through. I forced him to stay away from the office for nearly half a year, and he recovered. He also got his job done.

When massive hurricanes hit the southern regions of the nation in the fall of 2005, PAETEC wasn't affected by the one that hit Louisiana because we aren't located there, but Hurricane Rita plowed through the Florida panhandle and created a mess. Rich Stalder, our vice president of sales in Florida, had docked his boat on the canal side of his home, and when the storm was over it was sitting on his front yard—on the other side of his home. He was siphoning gas from the tank of his boat and pouring it into his car in order to get to work. All of our sales people were in a tight spot. We were looking at maybe a month before many of our potential customers there would be up and running—so there was no way to sell to anyone until the state recovered. This meant that many of our reps, who make the bulk of their income from commissions on sales, would be out of income for weeks. We recognized it would be a significant loss in income for them. To help bridge them through the dry spell, we calculated each one's average commission over the previous twelve months, and gave commissions to them for a month, even though they weren't able to earn the money just then.

I could describe hundreds of instances where we've allowed a family emergency to take precedence over the demands of the job, but one story will suffice. One of our best people, a couple years ago, was going through a very rough time at home. He and his wife were struggling to keep their marriage together. She was pregnant, but had health problems that were considered life-threatening. Childbirth was a major threat to her survival. She survived, but she was seriously ill after their child was born. So he had to be away from the office for weeks at a time. Over the course of three years, PAETEC allowed him close to four months of time off, to care for his wife and child.

One of our NOC managers remembers:

> We'd given him a couple months of time off, but we weren't sure how far we were sticking our necks out in doing this. So we met with our boss, who told us, "Give him what he needs, and when he's back, make sure he's working full steam." We did. Now, he'll give his blood for PAETEC. After his daughter was in good health for a year, we made him supervisor of a shift. He told me, "You can't believe how much I appreciate this. I came from a company that treated me like trash and I'd been there for years. I come here and you treat me as if I'm the most important person in the company." We had many teary-eyed moments with him when he was really down about what was happening at home. He'll volunteer for anything now. He was on vacation, camping with his kids recently in the Adirondacks and we called him and said we were short of people in the NOC. It wasn't an emergency, but he gave up his vacation time to come in and work because we were short. He'll stay around the clock. He'll be true blue PAETEC until the end. He says that all the time.

Another NOC manager has some reflections on why everyone approached this worker's situation this way:

> Here's someone who works for me, who obviously has a problem, and, just thinking like a human being, why would I not want to extend myself for one of my employees? When you get into a big corporate mindset, you'll think the opposite: it's all about how to get out of doing anything that might cost a little money. Now I have a guy who will run through a wall for me. After everything was said and done, his wife made it through the pregnancy, everything came back to normal, and he was very appreciative of life, but even more, he could

still provide for his wife and family. Ever since then, he's the one who, when I need a ticket looked at or a shift covered, I'm not done with the sentence before his hand goes up.

Here's the most important detail of this fellow's story. He'd been on the job only a week when all his problems began and he had to ask for time off. It would be hard to find another company anywhere that would invest so much in one worker who'd been with the company for only a week. But, for us, it was a crucial test of the PAETEC culture. We had to do what we'd always said we would do and put family first. It paid off for everyone.

Steve Hollis, senior technical service delivery consultant, tells a similar story, having been through a family crisis with cancer quite recently

In February, my wife had trouble swallowing. A GI scan showed a tumor in her throat. She was diagnosed in March with esophageal cancer. She went through ten hours of major surgery at the end of March with one of the leading two or three surgeons for this kind of operation and she made it through amazingly well. Since then she has required a lot of care at home, and Arunas said, "Take care of what's most important. Work will take care of you, and then you'll take care of work."

It isn't just Arunas. It's the caring culture here at PAETEC. You have to take care of family. Each and every one of us at PAETEC is part of a huge family that just grows and grows. Everybody looks out for one another. Whenever one of us or any of our children have issues, we all pull together. If you lose sight of that and something that goes wrong, you're reminded immediately what kind of a family we really have here. You can't find that anywhere else. I'm telling you, I've worked for a large company (AT&T) and I've worked for small companies. You don't have a culture like this anywhere. You don't have a family feeling that they're all in it for you, they are all fighting for you, they are all telling you, "I want to help you, I really mean it." They really mean it. They really want to help you. I can't tell you how that feels. I can't even express my gratitude for how everybody's treated me. Everybody gets involved in it—from your boss, to people you barely know who stop you in the hallway to say, "I heard about your wife. I can't believe it. Is there anything I can do? Let me know." These are people I don't even know! It's all just phenomenal. I have a huge loyalty to this company as a result of going through this ordeal.

It has been overwhelming. I'm going to the chemo and radiation with her, and I'm out of work at some point of every day ... my workday is shot. I don't get here until noon but—I'm still here. I probably should have taken her first week of chemo off. But, it is my little bit that I can do to feel like I'm giving back to come here when I can. I know that I need to give back to the company that has given back to me.

CARE ABOUT THE PEOPLE YOU FIRE, BUT FIRE THEM

It doesn't sound like brain surgery, but when you're talking about an organization of 1,300 people, keeping things at such a personal level isn't a small task. This caring culture can even extend beyond an employee's tenure here. It's been said that you don't know who your friends are until you're down and out, and, in a similar way, the best test for how much a company really cares about its people is how it treats the people it fires.

"I had to fire someone, a single mother, a woman we moved from Florida to New Jersey," remembers a former PAETEC executive. "She left on a good note. A single mom—I let her go, and she moved back into a family member's house. The house burnt down. Everything was lost. They had no insurance. Then I got an email from the head person in Florida, our head of operations. She said they'd collected five hundred dollars from Home Depot for this woman, and I wrote back and said the company would match it. With most companies, I think it would be, 'Gee, I feel bad, but, you know, she doesn't work for us anymore.'"

Firing people is as important as hiring. I call it pruning the rose bush. The way to get a rose to bloom properly is to prune with precision—to cut back the stems that aren't contributing enough to the plant. We're aware of that at PAETEC. To say we have a caring culture doesn't mean we're soft or too tolerant of people. In fact, we try to show people to the door after we have repeatedly tried to make them productive in their jobs or they just don't fit into our culture. The point is to keep treating them with respect and compassion, even *after they're gone*, when we have nothing to gain from it.

One of my officers has had to fire a number of people during her years in the business, but still finds it difficult. A few minutes after she let one of her people go around the Christmas holiday one year, I happened to walk by her office and saw her with her head lowered, crying. I went back to my office and told my administrative assistant to call her in.

"Did you have a tough morning? I heard you had a tough morning," I said.
"I had to let her go."
"Why?"

"She wasn't a good fit. She had a bad attitude."

"Was there another place to put her in the organization?" I asked. "Were there other opportunities? We could have moved her."

"I wouldn't pass her off to some other manager here. It wouldn't be right," she said. "I hate doing this stuff. Even if I know it's the right thing to do."

"Why?"

"This time of year. It's the holidays."

She started crying again.

"She didn't understand sales people are her *customers*. She didn't fit in at PAETEC. I don't want anyone on my team that's negative. She didn't realize what a great place this is and how fortunate she was to work here."

"That's why hiring is so critical," I said.

I'll offer one more example on this subject. We bought CampusLink, a company in Ann Arbor, that provided software for colleges and universities to operate their own telephone and data services. It was a very giddy period, 1999. CampusLink at that time had $13 million in annual sales, and we bought it for $43 million. Three years later, in 2002, the downturn was two years old. We weren't going to establish a beachhead in Detroit or Cleveland, as we'd thought. Students weren't using copper lines as expected, because they were switching to cell phones and Instant Messaging (IM). We had to shut CampusLink down and give people time to find new jobs.

The situation was this: we needed support from CampusLink from early in the year until October. We could have kept people motivated by not letting them know they would likely be laid off in the fall, and that would have given us the support we needed in that company. But we didn't lie to them. We told them early in the year what was going to happen. Dick Ottalagana and I flew out in January, and we gave these people nine months' notice to find jobs. It gave everyone more than enough time to find other work, prepare their families for a move, whatever they needed to do, while still earning a paycheck. They could leave in a matter of weeks, or stick around for the duration, and we even offered a bonus as an incentive to stay with the company until the end. No one gives anyone that much time to find new work. If they found jobs in four weeks, fine, they were free to go, obviously, but we gave them as much time as we possibly could to find new jobs. Anyone who wanted to relocate to Rochester or New Jersey was considered for a job. Twelve people wanted to. Afterward, people who found jobs elsewhere called and said, "If you ever decide to open an office in Detroit, call me."

Letting someone go is something that needs to be done right at any level of the company. Kathy Chapman, an ex-president for our West Coast region, recalls the painful process of coming to realize she had to leave the company and how it was handled:

When I joined PAETEC, I was made president of the Western region, so I was a regional president at the age of 32. We had never done business on the West Coast, and it was a different ball game on both the technical level and in terms of business contacts. Nobody knew who we were in California. Our cost structures were different. The woman who ran sales for me thought she should have been doing my job. "Why are you here?" was her attitude. I lacked experience in letting her spread her wings. It got worse and worse. I knew that was not the job I should have been in, but the pressure to do well for the company was intense. I felt very loyal to PAETEC. I kept wondering, "How did I get here?" I took it way too seriously. I had a four-hour commute to work many days a week. Other days I was sleeping at my sister's place in my nephew's bed with NFL sheets and glow-in-the-dark dinosaurs. My work schedule was so brutal, it didn't allow me to rejuvenate myself.

I put 60,000 miles on the car. My management style was very condescending, which wasn't how I meant for it to be. If you worked for me and didn't do what was right, you got blasted. I wasn't managing the way PAETEC manages people, but it was hard to see this from my perspective. Everyone else saw what was going on. Jack Baron became my boss, and I was demoted, and it was a relief. But the relationships had been forged already, and I couldn't undo what had been done with my people. I found a professional coach to help me, and the company was still really trying to support me. I paid $12,000 for a woman in San Francisco to interview all the key people. She gave me a three-inch thick binder—"Here's what's wrong with you and here are some articles to read,"—and she was on her merry way. Finally, Jack had to fly out and fire me. We both cried.

I don't think most people have the opportunity to fail. Failure is where you find out who you really are. If all you do is win, win, win, it gives you a sense of entitlement or ego and confidence that isn't founded in anything real.

I'm currently in an executive MBA program. I never want to feel as if I don't know how a business works or what's going on ever again. I want to parlay my experience with my knowledge and be the most productive person I can be. One thing I'd like to do is go back and teach a business class about how to become a good manager. At PAETEC people come first even when they fire you.

Jack certainly did that. He cared. One sales manager sent me flowers the next day. I look back now and think it was the best thing that ever happened to me.

RECOGNIZE, RECOGNIZE, RECOGNIZE

The single most important way to instill a caring culture is through recognition. We give people a level of personal attention—an awareness of who they are and what they do—that they can't find working anywhere else. I began this company with an employee recognition meeting. It was before I had employees. Before I had a *company* or even a building for holding the meeting. It was, to put it simply, before there was anything to recognize. I purchased 17 Mont Blanc pens, and I had them personalized, and I invited all the best people from ACC in a room at a local hotel. Ottalagana and I sent out invitations, and everyone we invited showed up. At the meeting, I said, "I want to thank all of you for what you've done for us. None of us would be standing here if it weren't for you. I'm giving you these pens as a symbol of my gratitude and respect for what you've contributed to my career." And then we started brainstorming about what we might all be able to do together again as a team. At the end of the meeting, Molly Korndoerfer said: "I'll wash windows. I'm in." And most of the rest followed, with greater or lesser degrees of Molly's enthusiastic leap of faith.

Recognition is what keeps an enterprise motivated and creative. If people know that you appreciate them and will reward them for going above and beyond the requirements of their job, they will do anything for you. But mostly if they simply know *you care enough to know who they are and what they are doing*, they will go to extremes to help the organization succeed. It sounds ridiculous, doesn't it? But it's true. You can never say *thank you* too much, and PAETEC says thank you in as many ways as we can think up. We conduct a variety of formal recognition and rewards programs.

One of our senior vice presidents recalls, in the early days, an award I gave her, before we had a name for it, after she'd been working day and night without a break, helping to get the company on its feet: "Arunas said, 'You and your family have my condo in Disney World, and you're going on vacation for five days. You've been working hard, your family misses you, you deserve it. You have five thousand to spend. It's the Old Key West resort. Right on Disney World property.' He was always very good that way. It was just between Arunas and me."

Strictly speaking, my gift to her was an informal hybrid of what has become our Maestro Award and our John Budney Award. Those are two formal ways we recognize people doing a good job. For example, senior officers give Maestro Awards to any employee at any time for a job well done. These awards can be substantial, a ticket to Disney World, a dinner for two, some stock options, you name it. We don't burden this program with any strict rules: any senior officer can give them out at will to any employee for an above-and-beyond effort. The John Budney Award is more substantial, and was named for one of our founding employees. The John Budney Award is for people consistently performing at a very high level. You get a Rolex watch, or a Ritz Carlton vacation, plus $1,000 to help with the tax impact. A $5,000 award. Only a couple dozen people have won it in our first eight years of doing business. My informal award of a trip to Florida, early on, would be a John Budney Award now.

One of our John Budney Award winners says: "It makes me feel good knowing only twenty-something people have it. John Budney nominated me. You get a watch or a trip. I picked the watch, because I'll have that forever. On the back it says 'From PAETEC.' No matter what you do in life you want to be recognized. In this company, no matter what you do, you *can* be recognized. I've called people out for recognition myself. I nominated a provisioner, a guy who provisioned circuits for us, for a Maestro Award. He orders the circuits from Verizon, which is very difficult. It's all part of being a relationship-type company. We form bonds and they last."

Some of our recognition programs require a little detective work, such as community service awards. People who provide something to the community don't usually brag about it. Similarly, the most valuable ideas for improvement don't usually show up in suggestion boxes, but happen during the course of everyday business. So the company tries to find out what employees have accomplished and writes up the improvement as if it had been submitted to a suggestion box. Employees are rewarded with cash and recognition: I have a regular Friday morning conference call with the entire company where I report on the latest news about the company and recognize people who have done an outstanding job—including creative contributors whose ideas have improved the way we do things.

Almost any way you recognize people is important, but I'll repeat what I said before: some of the most important acts of recognition are down in the subtle details of daily work. There are small ways you can make a good recognition event into something much, much better. It isn't about spending money, either. Sharon LaMantia explains how we do it in Irvine:

> We used to hold our quarterly recognitions at a place where there was a happy hour. Recently, we decided to hold it on the patio out behind the Irvine sales

office. For the first two quarters, we started at 4 p.m. and went until 7, on a Friday. We got people to come all the way through heavy traffic from Culver City and San Diego, which is seventy miles away, driving to Irvine to attend. We served beer, soft drinks, and snacks, and we handed out plaques, and everybody just hung around together in a very casual setting. Our participation for these shot up to a hundred percent. In a bar, people drift off, they are roaming around, watching the game, meeting other friends there. This way it's at the office, and it's kind of like standing around in the kitchen. Ties come off. People just relax. It's all PAETEC people, nobody else, so there's more good-natured heckling. You can hear everybody, see everybody, including all the back office folks. No music. Everybody is listening to everyone else and comments come flying across in the middle of sentences. You get a lot of loud teasing—brother-and-sister teasing. It's like everybody's gathering at Mom and Dad's house.

Yet the real glue that holds the organization together is even more subtle. It's the sort of recognition and care embedded in the way we operate and communicate from hour to hour and day to day. You spend a day with almost any manager in the company and you'll hear the sort of thing I'm talking about, whether it's one of our officers interjecting a few almost subliminal words into a conversation with one of her people on the West Coast—"You're so good. So good at what you do."—while strategizing how to get closer to our customers, or one of our VPs talking with her team about whether or not they should penalize people who aren't working up to their abilities or reward those who are doing an outstanding job. (They decide of course on the path of recognition: to catch people doing a *good* job.) It's the fabric of the way we work.

I'm known for writing notes, sending e-mails, sticking my head through the door, not just to thank someone for a job well done, but to simply say hello, ask how things are going. I try to be constantly aware of everyone. It's Management by Walking Around, the old Tom Peters axiom. One day I went into a VP's office and asked, "Who on your team is working the hardest?" She said Sarah Spencer had been working very hard on some problems we had with vendors. So I found a flower arrangement a vendor had sent us and I took it to Sarah's office and thanked her for how hard she'd been working. The VP told me later, "You can't measure the sort of impact that has." I didn't buy those flowers for Sarah. I'm a famous re-gifter. It was the fact that I found out about Sarah, found the flowers, took them to her, and thanked her. It's the gesture that counts, the awareness behind the gift. It's all about paying attention to people, as *people*, not as little nodes of productivity.

I skim every employee evaluation, digging into the person's job development goals and job issues. I look at a person's career plans, concerns, and see if there's a job at risk. I'll study the entire job appraisal for maybe ten percent of the workforce. I take them with me on JetBlue. It takes me three hours a month. I rip off pages—the form is six pages long—and file them. Some people say things counting on the fact that I'll read it. I can process forty reports over breakfast. One out of forty needs a response, usually. So I've taken care of thirty-nine.

For years, I've had my administrative assistant quiz me on the names of our employees. I'll assemble a slide show of the people who work in the company, memorize their names, and then put myself to the test. It's a small gesture, but it personalizes the business, and people are often surprised when I address them by name in the hall or at a meeting on the West Coast. I'm constantly spending travel time refreshing my memory on names and quick profiles of the people who work for me so that when I meet them, they will know I've cared enough to spend the time to familiarize myself with who they are and what they do.

Tracy Gaffney, the former manager of order processing in our Voorhees, New Jersey office, recalls:

> I remember when I was a senior account manager, we had several new offices opening all over the country. Many people were being hired at one time. Someone had asked us to come up for the first session of formalized training. Arunas walked into the room, and I was floored. He literally walked around and shook the hands of every one of twenty people. He knew the names of everyone he had met only once or twice before. That's impressive. To walk into a room and know which people you've met before.

> I'd only met him that once in the office at Woodcliff, and later when I had come up to the Rochester office, I was walking through the hallway, and he stopped me. I recognized him from the website, and he introduced himself. That makes a difference. That's the kind of guy he is. He's unassuming. When I was promoted, he wrote me a note: "Congratulations. Call me if you need anything that will help you in your career."

This kind of awareness and sensitivity to the need for recognition becomes a way of life at PAETEC: if the leadership makes it part of the way the company operates, everyone will get the idea and do the same. One former marketing employee remembers her first day at PAETEC, how everyone in the company took notice of her, not just the CEO:

During my interview process at PAETEC, I read through the literature and browsed the website. One of the messages that came through loud and clear was that PAETEC was all about people. I wondered if the PAETEC story was real or hype; but, hey, a job's a job, right?

In September 2003, I walked into the PAETEC foyer as a new employee. Josie, our receptionist, was there to greet me with a big smile. She welcomed me to PAETEC—using my first name! That was a shocker! As I rode the elevator up to the 3rd floor, there were no averted eyes, no "There is a stranger amongst us" aura. Everyone spoke to me and welcomed me to PAETEC. Another jolt.

My first day was over in the blink of an eye and there is only one memory that stands out clearly in my mind: throughout the day, everyone I met—at my cube, in the hall, at meetings, and, yes, even in the restroom—welcomed me to PAETEC and said this was the greatest company to work for and that I would love working here.

Remembering people's names and writing notes that show you know what is happening in a person's life and career are fairly superficial ways of making business personal. Yet these gestures have an enormous impact. Dozens of people hang on their walls, behind glass, a three-line note of appreciation from someone on the senior team. Not simply out of ego, but because it affirms that there's something human and personal that links the top of an organization with the middle and the bottom.

Recently, we created an award program to encourage advancement in our quality program. A person must undergo long hours of training to advance to higher certification in the Six Sigma and ISO 9000 system, and, as a motivator, we minted our own series of coins to signify each higher level of certification. At the start, an employee receives a wooden plaque with indentations—similar to those in the books a coin collector uses. With each advance to higher levels of Six Sigma attainment, we award a new coin. It adds a small element of fun by inspiring the fever a true collector feels in his quest.

In a similar vein, we've launched another program to recognize people who embody the principles of quality in their work. The idea is to collect all the letters in the word QUALITY, by being nominated for each letter by a manager. Every one of our 150 people with the title of manager or above can recognize an employee for the award—each manager is allowed to award one letter to one employee per fiscal quarter. Once a letter is awarded to an employee for exceptional devotion to quality, through unmatched service, a personalized solution, or simply exceptional work day in and day out, that person can go to a sales organization event. After collecting all the letters in the word, employees are awarded a free vacation trip with the winners of the President's Club in the sales organization: to Mexico, Napa Valley, Italy, wherever the trip happens to be that year.

Recently, I awarded my one letter for the quarter to Magen Jones. She's always happy to do her job, in the mailroom, at the reception desk, and now in human resources. She'd never gotten a letter before. I gave her a Q, to get her started. For that letter, she'll be able to attend a nice evening event and her name will be posted on a brass plate, under the letter Q, on a board, visible to everyone who enters the building, on our first floor. The next time she gets nominated, she'll get her U, and so on. Only higher executives are allowed to hand out the last letters. And only a handful of senior executives are allowed to award the final letter to any employee. That last one is the tough one. Managers actually have to pay attention and recognize what our people are doing, which is precisely how people show they care.

I'll let one of our officers say the final words on this subject, talking about how putting people first looks from the middle of the organization:

> People come up to me and say, "It's just your people. That's what's special about PAETEC. You have real entrepreneurial people." When Arunas heads a staff meeting, it's about people. When he goes through seven points in a management meeting, every point is about people. Everybody is concerned about my hours, how long I work every day. I've never believed in a company as much as I do PAETEC and that's why I work so hard. Of all the places I've worked, I've never worked where I've had such strong feelings for the people I work with. Everybody cares.

CHAPTER 2
Be a Family

EVERY YEAR WE TURN OUR HEADQUARTERS into a winter wonder-
land in mid-December, for the children of employees. Women dress in
red and green, floppy caps and curly-toed elf slippers. Santa, big and gray, the
perfect Santa, sits in one of the upholstered chairs outside our marketing
offices. Sleigh bells hang on ribbons around necks everywhere as Santa walks
around the building ringing his bell before a single child has arrived.

Soon, kids start to show up. Downstairs, you come in the door and see the
reception desk piled high with wrapped boxes and cotton snow, with a sign pointing
the way toward Santa's throne. The theme is the Polar Express, with train tracks, large
and black on sheets of paper. You go up the elevator and see the line of parents
and children weaving around the third floor, the semicircular glass walls that look
down onto the Network Operations Center (NOC), and the movie of Rudolph
the Red-Nosed Reindeer playing, over and over, on the jumbo wall monitor.

Somebody walks by Santa and says, "Look at him. Give him a beer."

Even early in the event, from upstairs, you can hear the first kids on the first
floor, their voices echoing up through the elevator shaft. Silver helium balloons
everywhere tug on their ribbons, tied to various objects around the office.

In the atrium, a sign says, "This Way to Snack Land."

Mary Lou Ignizio meets her grandkids, Jake and Alyssa, in the atrium
and tells them "Follow the tracks." They head up in the elevator and Alyssa
gets to press the button for the third floor. Jake makes a mental note to require of his grandmother the privilege of pressing the button the next time
they are in the elevator.

Upstairs with Santa, there are approximately 300 gift bags for the kids,
stacked like paper towels on shelves in a supermarket aisle.

An Elf asks the first child in line: "Would you like to sit in Santa's lap?"

"You want to ring those bells? Ring them and you might see a reindeer. Those are off Blitzen. Do you know who Blitzen is?" Santa asks.

The child isn't submitting to a quiz at the moment. He wants to put in his order.

"I want a truck."

"What color truck?"

"A red truck."

Red it is. Santa has several hours of this ahead of him. I played Santa once. Once was enough. I delegated the job. Jake and Alyssa move to the face painting station outside the elevators on the third floor. Jakes hides his face behind his grandmother. His sister gets a peppermint stick painted on her hand. At the elevator, Jake pushes button two. His sister complains, but it was his turn. She'll get to push two the next time. They move to the Music Room where a woman from the legal department is reading from storybooks. The kids sit dutifully on the floor and cross their legs.

Alyssa says, "I want to hear a story."

The woman tells a tale in which larger and larger animals climb into a white mitten. Finally, a huge bear tries to squeeze into the mitten with them and sneezes, blowing all of them out in all directions. Holiday lights hang all around the room, balloons are decorated with snowflakes, and a cardboard fireplace is seen behind the storyteller. A little tree sits in the corner, made of optic fiber so that the ends of the strands are lit up with different colors. The ends twinkle with light.

They get back into the elevator.

"Can I do two?" Alyssa asks. "He did two."

"Yes. Go ahead. Do two."

Jake reaches for one button, and Mary Lou blocks his hand. Alyssa gets to push the buttons this time. They go to the Wish List Room and write the things they want most. Alyssa writes Barbie and a computer. Jake writes truck and football. On their way out, they fold their lists and slip them into the little painted mailbox mounted beside the door. Nextel workers are wandering around with cell phones. "You want to talk to Santa's elves?" they ask and let the kids dial into Santa's workshop. Nextel workers are set up in their offices across town to answer the calls. (Nextel is a partner with PAETEC for providing service to customers.)

We go down to the first floor and hear keyboard music coming from the cafeteria. When we get there we see Jim Raub on keyboard and Jack Baron with a dark blue guitar, playing and singing carols. The tables are laid out with hundreds of cookies, water bottles and juice boxes for the kids, stacks of lyrics to the carols, so everyone can sing along. Jack is playing and singing, "Gloria in excelsis deo."

Jack's wife Lisa, an attorney, looks as if she's wise to the potential silliness of all this, but she's liking it anyway. She sings along when she can, supporting her husband who's up there putting his heart into the job. "Frosty!" she shouts. "Kick it up!"

"Away in a Manger" comes first. Then "Frosty."

Jim: "Now we're going to pick it up."

Jack looks at his wife, smiling. "We need more parents singing."

"Then you should be serving beer," Lisa says and grins.

Gift bags for the children are a big part of why an event like this works. We put a lot of thought into them. This sort of thing wouldn't be out of place in an MBA curriculum: how to assemble the kiddie gift bag. Any CEO who wants to build a company that feels like a family knows that on holidays it's all about the gift bag. These kids are going to be wandering around the building for a few hours, if all goes well. They need to be carrying something they can take home, something they can occupy themselves with when they're in line. Checklist: the little reindeer with the PAETEC scarf around the neck, the ten dollar gift certificate to Target, the kazoo, the small games for the drive home in the car, the Polar Bear craft kit, Santa chocolate, candy cane, popcorn ball, maybe half a dozen other treats. The caring is down in the details.

This is only one of the ways you build a close-knit family out of your team: make everyone's entire family welcome in the building. We not only encourage people to take whatever time they require to be with members of their family on important occasions, or during emergencies, but we encourage them to bring their kids to work, especially on holidays and for family-inclusive parties. We go to great lengths to make the children of our employees feel PAETEC is a special place they look forward to visiting.

The point is, the place needs to feel like home—not just to the people who work here, but to their children as well. When you do all these things, customers can feel it when they talk or meet with your people.

Holly Roedel Moore, who handles customer problems in our Voorhees office, says that she hears from customers the same thing over and over: "It's like a family in your place. Everybody works together. Can I come work there?"

FRIENDS BECOME FAMILY

At PAETEC, we begin as friends and, if we stick around long enough, we end up as family. The tone you find among brothers and sisters around a dinner table at home is the tone you'll find, more often than not, throughout the company. We try to make things fun, even during our busiest times, when things can seem on the verge of chaos. And that means working together out of love and respect, not because we're all following orders. The boundaries between work and home, personal and professional, are permeable. We're the same people at work as we are everywhere else in our lives.

Work is only one part of an employee's life. Most people work to make the money they need to do things that may matter more to them: family, school tuition, orthodontics, music lessons, golf, sailing, rebuilding a classic GTO, collecting comic books, tying flies. You name it: you have a day job so you can continue to design a cold fusion lab in your cellar or spend your Saturdays sewing Civil War uniforms you hope to auction on eBay. I won't laugh: my name is Arunas, and I collect chess sets from around the world. If it gives your life meaning, it matters. Our house is your house. But we expect, when you're here, to put as much heart into your work as you would if you owned the company. It's a fair trade.

The company's Balance Ratio—the balance between work and family—holds that if something personal urgently needs attention, a worker can simply take the time for it and figure out how to catch up, without a special request. The reason for absence doesn't need to be an emergency. It could be a child's school play. As long as the privilege isn't abused, people are encouraged to be absent from the job for the sake of something more important in their personal lives.

"When I first came here," one of our NOC workers says, "I was here for two days and then had to be gone for a week for my brother's wedding. No problem. I worked two days and had a week off, with pay."

I hold meetings out of town in such a way that people can catch a flight home in the afternoon, not at midnight. No one is required to come to a meeting if there's a Little League game to attend. Nobody gets yelled at for being absent. It's voluntary: your stake in what happens increases the more time you put in, but nobody is punished for devoting time to family. When you treat people that way, they are motivated to make up, later on, for what they've missed.

Dan Reinbold, who works out of his home near Cleveland, Ohio, recalls how the PAETEC emphasis on family had a big impact on his career:

My situation speaks to PAETEC's flexibility for the sake of families. I live in Cleveland, Ohio, and we don't have an office there. In my office, it's pretty much my dog and me. I'm on the road a lot. I have two kids who are now in their mid-teens. Back when they were twelve and ten, I said, "Hey, I'm going to be on the road a lot. And I won't be here some evenings." They were OK with that. If I had moved to Rochester that would have meant uprooting everybody. The word was, "Let's get Dan to Rochester." At the time, my wife and I thought, "Let's do it now before our son is more entrenched." So I walked into Arunas's office and sat down and said, "My wife said if we were going to come here, now is the time."

"What year is your son in?" he asked.

I told him.

"No. Stay there," he said.

It was as simple as that. It would have been better for the company, more efficient and easier, if I had moved to Rochester, but Arunas wouldn't have it. I was relieved. It would have been harder on everyone else. When I became VP of sales, my area was Buffalo and Rochester. I'm still living in Cleveland. I bought the house my wife grew up in, you see. It's a beautiful place, on a cul de sac and behind us is a park. To buy the house she grew up in, the house she still loved, shows how rooted we are here. My situation just shows to what lengths Arunas will go to preserve a family's happiness.

Another example of "Family First" comes from Mary O'Connell. She was with a Washington, DC law firm for nine years and decided she wanted to join the office of a public defender in a city where there was such an office, to help represent the indigent in court. She came to Rochester because, at the time, it was one of the few cities in New York State with a public defender's office, and then she went to work for Global Crossing. When she decided to join us, she was pleased by how flexible her hours could be so that she could be with her child.

One of the reasons I came here was that I had a young child and I wanted flexibility in my schedule. I'm an attorney and I'm only supposed to work 32 hours a week. Now I have flexibility. I can pick up my daughter if she's sick. I'm always conscious of trying to maintain and work as hard as everybody else. We work outside the traditional nine-to-five structure.

I had been in a law firm for nine years with no flexibility and Global Crossing had no flexibility. My daughter was an only child and my time with her was slipping away. When I started here I was home with her for one day every week, and it worked really well. She was home with me that day. People here were comfortable with the arrangement. I took my day off on Wednesdays. Unless we had an absolute emergency, people respected that time. When she went to school full time, I tried to be with her two afternoons a week. Now I spread it out throughout the whole week at a structured time every day. I'm home when she gets off the bus.

People really are committed to the company here. Some might think being flexible with a mother this way isn't fair to single people. But I'm taking a pay cut to do this and there are times when I work longer hours without overtime, so it's a balancing act.

In a caring culture like ours, this notion of family permeates all of the company's vital relationships. The company itself becomes an extended family whose members are workers, executives, customers and vendors. Customers are invited to have great influence over PAETEC's operation through regular meetings of advisory boards, and vendors are treated with respect. Internally, this atmosphere encourages a comfortable informality in the way people relate to each other, and rivalries are often friendly. Everyone is accessible to everyone else in this family.

I'm a family man. Family is, for me, the whole point of what I'm doing. For the 2003 Halloween celebration, everyone showed up in costume with his or her families. Children went trick-or-treating from door to door in the office. Customers brought *their* children as well. The best employee costume received a hundred stock options. So did the family whose child had the best children's costume. In the spring, the company conducts an egg hunt, and one of our employees will always dress up as the bunny. At almost any gathering where family is welcome, the children show up in droves. All of this pays off, ultimately, for customers, because it increases loyalty, commitment, and motivation, and those are the key ingredients for great customer service.

My philosophy on the balance between work and home is—make money, but enjoy your life. If you're putting in eighty hours a week, you're probably not going to be good for PAETEC or anyone else. If you can't get it done in fifty hours, something isn't right. Employees shouldn't give up their personal lives while making a living. As we grow, more and more people are giving up more of their personal time than I would like to see, and this is inevitable, until hiring catches up with workload, but it's something we try to avoid.

The emphasis on family has helped make PAETEC a family itself. To create a team of people who feel as if they really are members of a huge, blended clan, it helps to be able to start a company from scratch. In the early days, we were such a small team, all in the trenches together, that we bonded in ways that must be similar to what people who've been through a conflict together come to feel. The family spirit was incredibly strong.

When things are most difficult—as long as hopes run high—a caring culture forms all on its own. In our first few years, people here cared about their jobs and other workers in ways they'd never experienced before in a job. We believed in what we were doing, so it was much easier to care for one another. At the start, we were working without much furniture and one of our engineers, Bill Fitzpatrick, grilled a lunch for everyone—every day. He would dress in what he called his white zoot suit, a doctor's white smock, and took a break from wiring floor extensions in order to fire up some charcoal outside and grill dozens of hot dogs and hamburgers. It was the start of the free lunch program we kept going for years. It was back when the market was red hot, before the big dot-com implosion, when we dreamed that maybe everyone in the company would become a millionaire in two years. Needless to say, it's easy to feel like family in that kind of overheated environment, handing out beer to everyone on Friday afternoons. We were running about as lean as an organization can run, and yet, on a shoestring, we somehow managed to have moments that seemed extravagant. I recruited a college intern that first year to work from eleven to two and she did everyone's personal errands: dry cleaning, filling gas tanks in cars, anything that would take away from our effort at the office. She went around every day taking orders. When the Fitzpatrick grilling service stopped, we relied on a pizza shop, Arrivederci, for meatball submarine sandwiches. We all put on weight that first year. One of our VPs says, "I'll never eat another meatball sub in my life."

Eventually, some of us moved to the IOT Building. There were only a few offices in this place. Mostly mismatched cubicles and desks. Let me be more precise: it was nothing but a big white shell of a building, and a lot of echoing inside space. Offices accommodated four or five people, but everyone else sat in tiny cubes, with a telephone or a computer. But when people were working for PAETEC in that kind of situation, they were as emotionally close as people can get in a work environment. We had people there for two years.

That was roughly the time when, for one employee party, we booked an entire floor at a local hotel, The Hampton Inn. My wife Pam and I decided to make it a theme party. It was International Night, a cross between a costume party and a convention of U.N. delegates. Pam and I went as Japan. For Italy, John Piarulli built a gondola on wheels, a soapbox derby sort of vehicle. The gondola came with an oar, but the wheels gave it motion.

The "kamikaze shots" on our Japanese floating island display were, I think, directly responsible for the gondola race in the hallway, which got us banned from ever having a party in that hotel again. But while it lasted, as a group, we were as close as you could possibly get—and we were having as much fun as you can have without getting arrested afterward.

We've matured a lot since then, and our gatherings, more often than not, involve the children of our employees. People who joined us after those first four or five years don't have the same memories, and haven't been through the nightmares we faced, haven't tasted the joys of being part of a team where nearly everyone loved everyone else. Now it's more of a struggle to hold onto the sort of informality that makes our company feel like a family.

Here's a small example of what I mean. Not long ago, one VP was going to throw a little TGIF gathering for her staff and she said she'd make pomegranate martinis for everyone. But after a little reflection, she thought better of it. End of story. That says it all about what happens when you grow to a certain size: the cautionary corporate mindset begins to grow and everything becomes less spontaneous, less fun, less risky, and less like a family. Sometimes it feels like a losing struggle when you know you have to adopt that kind of prudent thinking in order to make the work environment comfortable for everyone you employ, but if you stick to your principles—that the most important asset you have is your people—the feeling can survive.

OFFER FAIR COMPENSATION

A caring culture means nothing if putting people first isn't translated into compensation. At PAETEC, executives understand that they can earn only so much more than the average employee: it's called the Fairness Ratio. Though this isn't a widely used term, it's an informal idea Dick Ottalagana and I used to talk about quite a bit when we were just starting out. It was a way we referred to our desire to close the economic gap between the highest and lowest ranks in the company. It's a mindset that grows out of our sense that we're not just a place to work, but a family—and it springs from the same sense of proportion as the Balance Ratio (the balance between work and family). It limits the gap between what people at the top of the organization earn compared to what the people on the front lines are earning. Though it isn't a hard and fast percentage, it's an inner barometer that governs more than just salary. It means there are no executive washrooms, company cars, no country club fees, or other perks often taken for granted by executives at most companies, where people at the top allow themselves to feel as if they are a different breed from everyone else working for them. This attitude has led to bankruptcy and prison for many executives in American business. Instead, we fly commercial. All of us. It's obviously a way of being smart about sharing the wealth, but it's also a matter of being fair, of not widening the emotional gap between how people at the top and bottom of the corporate ladder experience work at PAETEC. Sometimes we charter a jet if we can do a sales blitz together, because with enough people it's either cheaper or more effective than flying on a commercial airline.

If you show up late, you park down the hill, even if you're the CEO. I've trudged a long way through deep snow, getting in late after a breakfast meeting downtown, and people have told me they've looked out their windows and seen me walking to the front door through the blowing snow—and that an Arunas sighting like that actually makes a difference in how they feel about the entire company. I'll tell you why. We're on top of one of the windiest hills in all of upstate New York, and it's an arctic wind that blows through our parking lot. When people see the CEO braving frostbite to get into his own building, it makes an impression. It's only a gesture, but it helps level the culture.

From the first day of operation, PAETEC grants two weeks of vacation to everyone immediately. We give three weeks vacation at one year. Four weeks for five years. There's no carry over, because it's for a purpose. You need to de-stress and decompress. People here care so much, and work so hard, they can't keep doing it if they don't have down time outside of work. Vacations are nearly compulsory. Use the time or lose the time.

You didn't have to work six months to earn your time. And everyone gets stock options; everyone has a share of the company they work for.

Fair compensation is a genuine passion with us. It was at the top of my mind when I first met with our other founders, before we even knew what kind of company we were going to create. When we interview and hire, I keep in mind a lot of the things I learned more than a decade ago, while earning my MBA at the University of Rochester's Simon School. One of my professors put it this way: "It's all about how you set up the rules of the game." I learned how the sales force should be compensated—they get a share of all money we make from any customer they bring on board, for as long as that customer remains with us. How should wealth be shared? Through bonuses and stock options that build a sense of ownership and responsibility for the entire company's fate.

These rules of the game are designed to show, in the way that often means the most—in the form of money and benefits—that the company values people above all other considerations. Everyone comes to understand that assets, capital investment, cash flow, market share, and profits are all possible only because the employees bring with them to work a certain attitude and energy, nurtured by the sense that they themselves are the company's first priority—and the money they make will reflect that.

IF YOU CARE, YOU BUILD PROFIT INTO THE RULES

You can't compensate your people this way if you don't make a profit. Of course, everything in this book is ultimately about how to make a decent profit—in every sense of the word *decent*. But I'm talking about building a profit into the way you take on customers, from the first day. We invented a software program called Profit Assistant and loaded it onto every sales rep's laptop, and our reps use it to qualify customers. That's right. You have to *qualify* to become a PAETEC customer: we won't sign you up if you're too small or too large, basically, or if the price you want is simply too low for us, based on the cost of service at the time you sign up. We have to know, from Day One, that we're going to make a profit so that, from that point on, we know we can pay our people properly.

Bioworks is a small Rochester firm that provides ecologically safe methods for the agricultural industry to protect crops against pests. It's a worthy company, and we'd love to serve it, but we can't. Bioworks has tried three times to become our customer, and so far they still haven't made the cut. When we run the numbers through our Profit Assistant software, it just isn't profitable for us to take them on. So we keep turning Bioworks away, hoping either they'll get big enough to join us or our technological capabilities advance to the point where it finally makes sense to serve them. If we move to a next-generation technology it might make it affordable to bring them on as a customer. Until then, we have to keep turning them away.

We turn away customers for other reasons as well, reasons firmly rooted in the principle of caring culture. We don't expect our people to suffer from abusive or manipulative customers, and we'll cut a customer loose if it gets to the point where our employees have reached a dead-end with that customer.

Gary Eisenberger in our West Coast office, one of our best sales executives, will tell you about difficulties he had with one particular customer where the response from a CEO at any other company might have been quite a bit different.

We had one customer who ordered 22 T-1 lines in the same number of locations, which was incredibly inefficient. Something didn't smell right about it from the start. This was not the sort of thing PAETEC would have advised; in fact it was the opposite of what we would try to do, which is to consolidate services over fewer pipes. The guy making the decision, as it turned out, was getting kickbacks from the sales agent. It was not a good situation. When the deal was closed, it was understood that the traffic we'd provide was both inbound and outbound. When the time came for signatures on the contract, the customer said, "I want inbound only." We told him we would make no money on inbound. We'd have never agreed to the deal on that basis. But we went ahead with it, and all along, through the installation, he harassed us, berating people in my office, calling me up and telling me he was going to call my CEO.

As it turned out, he'd signed a contract he didn't have the authority to sign. He was backed into a corner and trying to make the deal fall apart by blaming us for problems. When he threatened to call my CEO, I told him I would handle it. Before I could do that, I got a call from Arunas. He was very calm. "What's up? I just got this raging voicemail from this guy." I told him it was a bad deal and a bad customer. He said to me pointblank, "Is this customer going to treat our people the way he treated me on the phone?" I said yes. "Are we making money from the deal?" No. "Do we need him for any other reason?" No. "I'll call and cancel our contract," he said. And he did. The next time I called that customer, three weeks later, I discovered they'd fired the buyer. "Can we get you back in here?" they asked. I told them, "'We should do one big pipe, not 22 of them." They said to come in and talk about it. This time we did it right.

As you progress through the principles of this book, you'll begin to see how each individual principle interlocks with and bolsters all the others; in some ways, you can't practice any single principle effectively without having all the others in place. Something as specific and technical as the Profit Assistant, for example, becomes essential to giving a large segment of our employees—our sales force—an attitude of ownership, which grows from caring. It empowers people to make decisions on the spot, in a customer's office, which otherwise would have to be made days or weeks later by committee, or at least in the office of a sales manager or VP. It gives people on the front lines the ability to make decisions often reserved for officers back at central command—as long as the software gives them the green light. (And yet profit can take a back seat if a customer is simply not someone who is aligned with our cultural values. We'll take a hit, financially, for the sake of our values.)

Mario DeRiggi, executive vice president of direct sales, describes how much more power he and his people have to make significant decisions at PAETEC in comparison with the atmosphere at larger telecoms, because of both our compensation policies and the way we ensure a profit through the Profit Assistant. The sales rep knows that he or she isn't going to make money, long-term, from an unprofitable customer, and so has a psychological hedge against cutting corners for a quick commission. DeRiggi explains:

> When I worked at AT&T, it was bureaucratic hell. They put in as many levels on the organizational chart as they could in order to stop people from making decisions. Here it's completely different. They tell you, "Here are the keys to get it done, and if you make a bad decision, OK, but don't do it again." Everybody talks about empowerment, but it's how you define it. Early on, we push a lot of decision-making out in the field. We want the people closest to the customer to be making these decisions. In sales, the Profit Assistant empowers people, because they know what they can and can't do with the customer. You want to talk empowerment without over-empowering people: it's a way of putting structure around the decision-making process. So you know if it's a good or bad decision-making process. It's only a tool, you can go around it, voluntarily, with sign-offs, if by doing that it's a deal that will be good in the long run.

> Recently, we saw a deal that looked profitable at a price the customer loved, the way we'd laid it out. But after reviewing it, with our software, we realized we really were losing money on it. This was tough. We were signing the customer up an hour later and we had to tell them we were wrong, and would have to raise their costs 15 percent. We got on the phone to the customer and said we had made a mistake. They weren't happy. The Profit Assistant software brought out the fact that we'd made this mistake, and we looked bad. It looked like bait and switch. But it saved us. We educated them on the process. We told them we'd show them everything about it. "Hey, sign a non-disclosure we'll show you the tool and how we made the mistake." We screwed up, but they're signing with us anyway. And we'll make money. Without the tool, we wouldn't be making that money.

PAETEC is old-school when it comes to money. We want to sell something and have a good margin on it, not play the system as so many other companies were doing when we were just starting up in 1998. We don't venture into a line of business and flip it in two months for a profit. Old school is being fair in how you view the business and the people: in a caring culture you don't play games with people and you don't play games with numbers. We want everyone taking personal responsibility not just for the current quarter, but for our entire future. My brother Al, our senior vice president for finance, takes the issue of integrity very personally:

> We want everyone, in every action, thinking, "This is our money we're spending, not somebody else's." You should have that responsibility anyway. I, as controller, have a defined job. I have certain fiduciary responsibilities. Post-Enron I'm now a little more focused than I might care to be. On integrity, I've said it to Arunas, this place is going to be clean. You know what? On TV, who's the first person who gets led away in cuffs? It's not the CEO or the CFO. It's the poor controller. I'm the first one who gets led away in handcuffs. The third one down. That's not going to happen. This place is clean. People got away from principles-based accounting a few years ago. They were focused on rules-based accounting, which devolved into a game of playing those rules, figuring out ways of tricking a profit out of the system without technically breaking the rules. The thinking was, can we make this transaction fit this set of rules? Even now, Post-Enron, everybody talks a good game but when the rubber meets the road some people *still* try to play the rules game. Principles-based accounting: we can't get there fast enough.

All of this translates into: "Let's make an honest profit instead of tricking good numbers out of the system." This kind of conservative approach doesn't mean we scrimped or cut corners in our capital investments. Our fixed assets are, in some cases, some of the priciest. Take computer storage, for example. We simply started out buying the most expensive storage. But we were thinking long-term profitability: it wasn't only the best equipment, it was also the most scalable. We planned on growing rapidly, and we did, and in the long run, we saved money, because we'd spent a lot up front. In other words, we made the most responsible decision for the sake of gradual, steady growth and long-term stability, unlike some of the dot-coms in those final years of the '90s whose burn rate was suicidal, simply because they believed that their only task was to be the first and biggest name in their industry on the Web, as if growing revenue was all that mattered. That first-mover advantage didn't carry the day, though. The top line rarely mattered more than the bottom. It's all a way of taking responsibility for the long-term growth of a company, and the long-term careers of the people within it. It's about caring.

Compensation isn't quite as simple as figuring out whether or not a deal will be profitable. Keeping people happy, financially, is a very difficult task, fraught with deeply personal issues of self-esteem and motivation. It's not psychobabble to say that people confuse money and love. If you put a workforce on a wage freeze and tell them everyone's going without raises for a business cycle or two, everyone feels rejected, underappreciated, and hurt. And yet if you get out and walk around and test the waters and try to address the issue personally, you will step into a world of such emotional complexity that it's hard to make a choice that works for everyone. In every decision you make as a leader, you must ask yourself: "Who am I missing? Who am I leaving out? Who is going to feel overlooked?" I'm not afraid to drop in on people and ask whether they're happy. I do it all the time. It's impossible to do this with everyone, spread out across the map. But you can't stop trying. Often, once you get past a certain size and level of financial stability, if people aren't happy, you can find out why by following the money.

I recently stuck my head through the door of one of my best mid-level executives, a woman who has been with us almost from the beginning, and I asked, "Are you happy?"

I like to get right to the point.

I hadn't spoken with her in a while, but I'd recently read through her regular job appraisal—where employees are encouraged to say if there are any issues they're concerned about—and she'd written that she wanted to see her compensation aligned with her colleagues. This comment was hidden away: she'd put it in a section she knew I usually didn't read. She wanted her boss to see the comment, but she didn't want me to think she was complaining. Even so, I spotted it, because her job satisfaction was, in my view, crucial for the company at that point.

"I love my job, but I'm not happy," she said.

"You're not?"

"No."

"Why?" I asked.

"I'm not going to tell you when you are standing there in the doorway," she said.

I shut the door and sat down.

"I had a conversation with my boss about how I didn't think I was compensated the way I should be. He told me he'd given me a nice bump the year before, that I was just promoted, and that without my taking on extra responsibility he would have a difficult time giving me an increase. If I took on more responsibility, yadda, yadda, whatever. And that I was aligned with one of my colleagues. Later, the finance department sent me our budgets, our head count budgets, and they sent me his by mistake, included with mine. I saw his salary, which was quite a bit more than mine, so I was not happy. I've taken on a lot of responsibility. It isn't as if I just do my job. I get involved in other things. I'm sorry, but I know what some of my colleagues make, and I think I contribute as much if not more than they do."

"I agree. You do. How much should you make?" I asked.

She told me.

"OK. I'll give it to you. You've got it," I said.

When the increase went through, I returned and asked her, "Do you feel valued?"

"Yes," she said.

"You know how much we appreciate you," I said.

"Yes."

"OK. Good. See ya."

This was a good example of how personal the act of doing business gets at PAETEC. It also illustrates how difficult it can be to make it personal. After all this took place, the story got out and made the rounds and my mid-level executive felt a small wave of resentment for what looked to other people like favoritism, as if I'd singled her out and rewarded her for *unjustifiably* personal reasons. I also took some heat from her boss who felt I went around him. He was right, I should have consulted with him. I'd treated her in a personal way, the way that I treat everyone who works for me—with personal concern about that individual's well-being. But, as this situation indicated, it can be misconstrued—and the simple fact that the many aren't as valuable to an organization as the few highest performers will be felt as fundamentally unfair. It is, but only in the sense that life as a whole is unfair, doling out brilliance or good looks or a great personality or inherited wealth to a select few. Compensation can't correct the unfairness of how human excellence isn't equally distributed. But the attempt to treat people with as much personal care as possible, especially in the area of compensation, will always pay off, especially if it starts at the top and works its way down the organizational ladder so that anyone who leads will follow your example.

DELAY YOUR GRATIFICATION

One of the most important ways we have attempted to fairly compensate our people, from the beginning, was the care we took in how we raised money to get started and keep operating. A caring culture means doing everything possible to include everyone in the financial rewards of success, and that means refusing to take shortcuts that would water down that shared wealth. When we raised our initial $134 million in private equity, I made sure that the deal wouldn't dilute the share value for the people who came onboard hoping that their options would some day help put one of their children through college or provide the down payment on a home. Jeff Burke tells a story about how quickly we went through that war chest and how we reached a point where we could see, at our burn rate, the possibility of bankruptcy, simply because we might not be able to build our markets quickly enough to offset the cost of doing business from day to day.

In 2001 and 2002, we were getting worried. We'd already gotten the big infusion of capital, $134 million, and we'd burned through three quarters of it just getting ourselves up and running. Our operating costs were eating it up, not capital expenditures. Another venture capital group offered us $50 million in cash, and we could see that without it, at the rate we'd been going, we might file for bankruptcy within a year. That new cash would have been a safety net.

I was an outside director on the board at the time. Arunas asked the board what we wanted to do. He didn't want to accept the money because it would have seriously diluted the share value for everyone in the company. He felt we had a credible leadership team with a proven track record, and wanted to take the risk. Arunas said, "We can do this." Keith Wilson, our CFO, went along with that stance. We had the talent, we had the vision, we could make it work. But there were no guarantees. It was an enormously exhilarating moment. Anything could have turned against us. Technology might have taken a new turn. The competition might have suddenly heated up.

But we believed we could win. The hair stands up on my neck just talking about it. It was a turning point for the company. The three independent directors on the board were the swing votes, really, and I was one of them. "Management has delivered on what they say," I said. "I vote for management." And we decided not to take the money. At that point, we knew we had to succeed, we actually had to start making a real profit, or we were done for. It was one of those moments that affirmed and defined the character of the company, and it was very much against the grain of what had happened in the dot-com era—that attitude of taking the money and running.

Later, a large telecom wanted to buy the company. But it was for stock, not cash. What a mistake that would have been. Arunas said, "No way." He didn't think the deal reflected the true value of the company. It was ultimately the same issue: "This is what we're worth, this is what I want people to get for their investment here."

It came down to three words: "We're not settling."

This became a mantra that applies to everything we do. The idea that we won't settle for less than the best. Arunas says it in a variety of situations. He keeps telling people, do not settle. Don't settle for half-baked solutions. Don't settle when you hire somebody. Get it right in the basics. It's his response every time we've built up to an IPO. Don't settle for less than the shares are worth.

Jeff is right about why we've backed down from going public several times, when the move could enrich many of us at the top of the organization. This same sense of obligation to protect the value of what our people have invested in this company has governed the timing of an IPO and how we've backed down from it every time we've gotten close to going public. I've assumed I would take the company public since I sat down and scribbled ideas on the back of a napkin with Dick Ottalagana. The question has always been: "When?" We were ready to take the leap just as the market plummeted and the dot-com bubble burst toward the end of 2000. Since then, we've been waiting—growing steadily, expanding our network footprint and the markets we serve—for exactly the right balance between our own profitability and a rising stock market. It's been a much longer wait than any of us anticipated.

We didn't *have* to wait. We could have gone public almost any time during the past four or five years, and those of us with the largest stake in this company—the owners with the largest blocks of private shares—would have made quite a bit of money. But, comparatively, everyone else at PAETEC would have lost. Not in an absolute sense. They could have cashed in their stock options and made some money, but nothing close to what they have all come to expect, and deserve, based on our remarkable rise. Our people haven't been hoping to retire to a tropical island when we go public. But the money many of them could make, if the IPO were timed properly, would help pay down a mortgage or put children through college.

Also, if we were to go public at a low share price, our private equity investors at Blackstone and Madison Dearborn might have gained control. The lower the initial share price, on the first day of trading, the greater their share of ownership in the company would become. All of which would dilute the value of stock held by a PAETEC employee or smaller PAETEC investor. When it's in the best interest of everyone at the company is when PAETEC will go public. It couldn't be any other way in a caring culture.

When you give people ownership, literally, through stock options, then it's a bit easier to expect them to have an emotional stake in the company's welfare. This is what we expect from everyone. The message from the top is: *I'm a leader of this company, so, yes, I may get richer if we go public, but you will get proportionally wealthier along with me. So we expect you to act like the owner you are, as well.*

When we started, in 1998, executives purchased over four million shares for themselves. Top officers received no options. The options went to employees. The receptionist, for example, got a thousand her first year, and a thousand each year—as did everyone else—for two years after that. After several years we changed the policy so that directors, and those above them, get two hundred options annually and all other employees receive a hundred. In the future, all employees may not get an automatic annual grant. Options are awarded when someone starts with the company and then when that person is promoted. If an employee achieves a certain level over time he or she will get a certain level of ownership. A lot of companies are doing away with options, but we aren't. How many options you get will be based on the level of your entry into the company, and how you succeed over time. ·

We didn't simply distribute ownership down the organizational chart, we also limited ownership among our independent investors. If you want to put employees first, you can't give up the reins to absentee owners. There has to be autonomy at the top, with power in the hands of someone who walks the halls every day: a committee of investors can't run the company. We founded this company with people who had already spent a decade or more at ACC or Rochester Telephone. We knew the business, and we knew how to lead this company. I'd been president at ACC, and I helped lead it through a period of strong growth into the '90s.

The core team invested the millions we made on the sale of ACC into the new company, yet we raised another $350 million from angel investors, venture capitalists, and banks. Though we sold a majority of shares in the company—management and employees still hold 30 percent of the company, with 18 percent in the hands of management—I held firmly to the power to veto anything my investors pressed me to do. I was able to do that because I had a track record with my team. When Blackstone and Madison Dearborn and other venture capitalists and banks anted up, they were investing in a known quantity. We've depended on their money, but we all needed to be part of the decision-making. Too many telecom companies were forced to do things that their management teams recommended against. That was their downfall; we protected ourselves.

CHAPTER 3
Commit to Community

INTO WHAT WE CALL PAETEC Plaza, which is our large, two-story atrium in corporate headquarters, I've invited members of The Women's Foundation of Genesee Valley for a cocktail hour. We've attracted more than a hundred people, and they are mixing on the ground floor, on the mezzanine, and up to the third floor. A musician in the corner in a tuxedo is playing an electronic keyboard with an amplifier, some lounge/karaoke version of an old '80s song.

All day, I was wearing an old striped shirt, open at the collar, a pair of worn trousers, and blond-colored socks. Before the party, I ducked out of a meeting early, drove home, shaved, put on a dark suit, blue tie, white shirt, and a pair of nice shoes. When I returned, the place was filling with women of all ages: retired women, young single mothers, middle-aged, engaged women with thirty other organizations begging for their attention, CEOs and housewives. All of them are dressed in business attire, mingling, drinking, eating from tables where chefs are quick-frying a little bowtie pasta or tortellini with Alfredo sauce and asking what they want in their dishes—"Shrimp? Broccoli? Artichoke hearts? Greek olives? A little of everything?" The women are loving it. As I go from one group to another, I hear them—some of whom have brought their children along, and children are always welcome at PAETEC, any hour of the day—sharing their stories of single motherhood, all of them on their own, but successful, a loose confederation of independent, female resolve.

Once all these women were assembled and relaxed, I came down the winding staircase from the second floor. I felt a little like a performance artist: I'd assembled all this partly for the pleasure of letting the community use our beautiful facility, but partly for the *art* of what was actually happening—the networking going on, the business, one way or another, we'll draw in as a result of something that happens here tonight. You set up all the elements of an event like this and it just happens on its own. We'll let these professional women do whatever they want to do in this building for the course of the evening, and, one day, eventually, we'll see the impact on our bottom line.

It's just a free party for a worthwhile women's group, most of whom come up to me and gush with appreciation for PAETEC's generosity. But the reality is, an event like this costs us maybe $8,000. Over the past two years, we've held 20 of these events, all of them free to any non-profit that wants to hold the event here. This group helps single women, single mothers, with grants, training, and support. The group has compiled "appreciation books" in which people submit photographs and write tributes for women in their lives, and these are on display.

On average, of the dozens of similar community events we've sponsored, we bring in $34,000 of new annual revenue from each one. It's inevitable. That's a 300 percent return on our investment. What I love is the indirect, effortless, fun way in which you can make that new business happen by simply drinking wine and joking with people and accepting their affection.

The women have invited parents, brothers, other family members, and a couple dozen children. On the third floor, when I go up, someone's little girl is sitting in my swivel chair, at my desk, and a small cadre of elderly ladies sit around my conference table with their tortellini and white wine. I'm fully prepared for the children because I stockpile fluffy PAETEC dogs to hand out. We buy them in bulk. They wear a PAETEC logo and have an emotional feel unlike that of a coffee mug, and at eight bucks a pop, they don't cost us more than a box of mugs. The children love these dogs, and that's partly the point of handing them out, but they also take them home where the dogs hang around, whispering *PAETEC, PAETEC, PAETEC.*

When I spot the little girl sitting in my chair, I go away, grab a fluffy dog, carry it into my office and hand it to her. She hugs it and all ten women in the room—on the couches, around the conference table, at the desk, with their wine and their pasta, go "Aaaaahhhh." My day is complete. The atmosphere here is wholesome, homey, and genuinely charming. Children wander everywhere, clutching their new pets, gazing up at the huge screen in the NOC, as I offer to take anyone who's interested on an impromptu tour of our facility. This is, above all else, a *safe* place to be. We want it to feel like the next best thing to home.

Outside, another woman in her mid-thirties, buttonholes me:

"I want to compliment you on all the artwork in this place. It's wonderful."

I smile on behalf of my sister, Jolanda, who designed the interior.

"They're mostly from online auctions. Really cheap," I point out, always quick to look cheap on behalf of investors and employees.

She grins, not quite sure what to say.

"Well, in Disney World all the buildings aren't really buildings, either," I explain.

She laughs.

Children rush by, and someone calls out "Honey, come here!"

Chris Wilson comes up and introduces herself to me. She's the daughter of Joseph Wilson, the early, legendary leader of Xerox Corp. when it was first establishing its markets. You can tell she thinks maybe I might be a throwback to those times when corporate leaders were human beings who knew that business was mostly about people and only secondarily about money.

"It's wonderful the success you've had," she says.

"The way I like to put it is, we have a good home in a really bad neighborhood. Telecom is a tough place to be, but we've managed to rise above it."

Another woman asks: "What are your revenues, can you tell me?"

I rattle off the numbers. We're an open book.

ALWAYS GO FOR THE TRIPLE PLAY

A caring culture means opening your doors and your treasury to the community. The motto at PAETEC is to go for the triple play: do the right thing, make money *and* feel good about it. Community service is one of the most effective and subtle ways to hit a triple play. It's a huge imperative at PAETEC, not simply because it's the right thing to do, but also because it is one of the best sources of new business. We have a program that grants all our managers a discretionary budget which they can draw from in order to support programs for the benefit of their community annually. Each director gets $2,500 to spend. Each vice president has $5,000, each senior VP $10,000, and each executive VP $25,000. They are allowed to give the money, which PAETEC has budgeted for community service, to a program where they, or a member of their family, have donated money or time, or have workers who do so. It not only creates a bond between PAETEC and the various communities we serve, it encourages our people to get personally involved in community programs. Over 65 percent of our managers sit on not-for-profit boards. Many others are involved as volunteers or committee members. Rather than have a central officer in charge of community affairs to decide which organizations are worthy, we decentralize this process around the country and build enormous commitment on an individual level. It makes our people passionate about the life of their communities.

But our sense of community involvement is about more than donating money or time. It's intimately interwoven into the way we do business. In only one of dozens of examples, PAETEC participated in a program that allowed us to hire young area people as interns. Sixty-four students applied, with sixteen finalists, and four winners. Of those winners, we brought on board three of the most motivated workers we could have found anywhere, for only $10 an hour. We hired three top-notch high school students, Young Women of Distinction, who also received a scholarship for college.

In this program, everyone benefits. These interns had great writing skills. They knew Excel spreadsheets as well as anyone in the controller's office. And they would be starting college at some of the best universities in the country, becoming, four years later, prime candidates for hire. Yet the benefits didn't stop there. Channel 13, one of the local news stations, produced a free promotional video about PAETEC for debut at the ceremony honoring the students, a video which is useful to PAETEC in marketing and public relations. The video and the related celebrations, some of which were held at PAETEC's headquarters, will bring to the company a minimum of two new customers through the networking that goes on.

By the same principle, PAETEC encourages all managers to serve on not-for-profit boards. It's not required but it's encouraged. The community benefits and PAETEC benefits by the networking with other business leaders. We encourage our people to volunteer. Again, it's not required. We participate actively in United Way and, as part of our Chairman's Gold Club, host a cocktail party for our contributors who gave at least one percent of their salaries to the United Way.

In simple terms, you can't be a taker without being a giver. You can't live in a community and profit from it without giving back somehow. It's a triple play: by giving, eventually PAETEC will get many things in return, and everyone feels good about it.

PAETEC won the Business Ethics Award two years ago from the Rochester chapter of the Society of Financial Professionals. In 2005, we won the national award from the same group. A Kansas City construction firm won the award for small businesses. We won for medium-sized firms. And Whirlpool won the award for large corporations. To accept the award, Jack Baron and I went to Phoenix. We saw the opportunity to do a number of different things with this trip, so I brought along my 14-year-old son and 12-year-old daughter. Jack brought his 15-year-old daughter.

We took our kids whitewater rafting and horseback riding in the desert. We also brought them to the three-hour awards presentation. During our time at the event, we got to know people in the organization, and we decided to put together a marketing program in partnership with the society. It will offer a referral opportunity to all members of the organization—any member who refers a new customer to us will receive a cash award from PAETEC. On top of that, during the flight to Phoenix and the flight back, Jack and I spent ten hours brainstorming new product development. So it was more of a grand slam than a triple play. We found four ways to capitalize on our award: we bonded with our children, we accepted an award which will serve as good public relations for us, we hatched a plan to generate new business through our association with the society, and I found time to meet for ten undisturbed hours with one of my top people to come up with new ideas for product development. To spend that much time spitballing ideas and thinking about the future would have felt almost criminal back in the office with the schedule both of us have.

When we won the Community Visionary Award in Rochester from an organization called Seniors First—which provides essential services to seniors living independently at home who need help with daily activities—we invited fifty of our people to attend the black-tie award gala. They were both managers and non-managerial people from many different departments: operations, finance, and marketing. We identified the most teamwork-oriented and community-involved people, and we brought them to the event, courtesy of the company. There was a fine dinner for everyone and a great swing band. Afterwards we invited everyone to gather in one of our favorite suburban night spots to extend the evening. It was a triple play, because we got the community recognition, we offered deserving employees a special evening of entertainment, and we also donated to the organization's Alzheimer's care program by buying tables at the event.

One year I met a high school graduate from a Rochester suburb at an awards event held by the Association of Fundraising Professionals. This student was being recognized as the top student volunteer in the Rochester area. I was already interested in hiring him, given the nature of the award, before I met him. His table was near ours in the ballroom at the Hyatt hotel, so I went over and introduced myself and said to his parents, "I'm impressed with your son." I told him, "If you want a summer internship, I'll hire you." He said, "Yes. I'd be interested."

So we brought him in the next summer. I gave him a special and appropriate mission: he acted as a sleuth, trying to uncover the names of all the people in our company who had done things for their communities but hadn't talked about it. I wanted him to identify the most worthy people of all—those who do good things for others but secretly, without making a show of it. He found the names of 172 people we'd overlooked whenever we recognized those who have done things for their community. One of our design experts put up a website where we posted photographs of all these people: you can click on their photographs and see descriptions of their contributions. So out of our support for the Association of Fundraising Professionals we achieved another triple play: we found a young man who will make a fine employee if we can snag him once he's out of college, we discovered a way to recognize people who might have otherwise continued to be overlooked, and we created a highly visible recognition program that shows others, who might not be quite as community-oriented, what a great idea it would be to serve their community.

I'm on the presidential advisory board of our local Monroe Community College. One way we're making a triple play in partnership with MCC is through a course especially designed so that all the students attending the class are assigned to work at PAETEC for the semester. So twenty students, with initiative and aptitude, come to us to work on a "real world project" for their spring semester. They get credit for it, and we get the manpower. Two of the students earn an internship with us.

For our first group, we focused on customer service. Students called our active customers to locate the person in charge of solving telecommunications problems when they occur—the person we need to deal with directly when something goes down. Before this class was assigned to the task, we simply didn't have the names or phone numbers of these people in our databases. The students were able to make these calls at any hour, from any location. They have Internet access and were easily trained on how to input into our system the information they gathered. It was a win for us, for MCC, and for the students.

PART TWO: OPEN COMMUNICATION

CHAPTER 4
Keep the Door Open and Listen

T.J. KULPA WORKS ON OUR MARKETING TEAM, helping to guide new product development. It's a trial by fire, because he's working completely against the grain of what made us strong for the first five years. This requires some explaining.

T.J. stands, along with everyone else in marketing, on the fault line that divides the way PAETEC used to approach its market and the way we will be doing it from now on. During our start-up years, sales drove everything we did. Imagine it this way. We sent our reps out to ask customers what they wanted from us, and then they came back and said something like this: "They've given us three weeks to make a widget, and they want us to do it while we're standing on our hands." We would assemble a team—visualize all of us coming into the first meeting, walking on our hands, and you have some idea how eager we were to please—and, *two* weeks later, we delivered a widget. A week early. This is partly how we grew so quickly: we never said no, and we exceeded expectations. Whatever a customer wanted, a customer got—even if it wasn't always the best thing for *us*. Our customers ran the company.

We'd like to think of our customers as the ultimate authority over what we do. But things have changed. We actually say no, now and then, just as we say no to some potential clients every time they come to us to see if they can qualify to be our customer. We do it now because we know a little better what's best for all of us, especially the customer. There are some customers we just can't serve profitably—our Profit Assistant software has told us that from the start. And there are some products and services we can't afford to provide. But this new hesitancy to do *anything* we're asked is about the future. We see the future, and it's the Internet. In fact, the future is already here. It's our job to help our customers wake up to it. Our customers are beginning to realize they can converge voice and data over the Internet and pay a fraction of what they've had to pay in the past, and we're positioning ourselves to help them migrate from the old way to the new. If we don't keep up, of course, we're sunk.

We plan to be ahead of the curve, not running to catch up, and it's because of people like T.J. He's intimately involved in this transition, and he thinks strategically about all the implications of this, on behalf of the company and the customer. Every day, figuratively speaking, he takes his passion for the business outside our walls and fights the good fight on behalf of the customer.

He has about twenty years in the business, with stints at ACC and MCI (then briefly WorldCom), before joining us in 2001. As soon as he signed up, he knew he was in a company different from the other places he'd worked. This is how he describes it: "It was amazing that in a company this size, you had so much exposure to how things were designed and how processes were put together. I saw the company coming together, and saw processes I'd never seen. That was a plus and a minus. In a larger company, if you make a mistake, you might not be held accountable for it. If something goes wrong here, you're the goat. Luckily, I haven't been the goat all that often. If you're right at PAETEC, it pays off. It gets noticed. You can be the go-to person if something else comes in as a result of it."

T.J. recognized the challenges ahead of us, and he realized that his job, in marketing, would be to push back against the momentum of the entire organization: in other words, choose to be the goat day in and day out, for the good of everyone else who didn't see how and why the company and the market were going to change. At first, a lot of PAETEC people didn't think we even needed a marketing group. Many of them, even now, don't think we do. They are still living in the days when, as T.J. describes it, "We were a company that catered to the customer's needs to the point where we'd do anything they wanted. If they need plumbing done, we'd send somebody out to get some pipe. Now it's completely different."

Now, T.J. and his team assess where the market is going, how new technology is going to change it, and what customers are going to need—whether they realize it or not—rather than what customers believe they'll need. He works hard to persuade people both inside and outside the organization that the future requires some unfamiliar kinds of adaptation. A company we serve may not realize how it needs to adapt, and it's our job to show how this can be done. For example, PAETEC is staking a big part of its future on Voice over Internet Protocol (VoIP), or, traditional voice service transported digitally over the Internet rather than with the usual analog phone lines. To achieve adequate quality of service for VoIP, customers are best served by an MPLS system—a multi-protocol label switching network. In layman's terms, it's a network that lets you attach many descriptive labels to a bundle of data so that the bundle can be quickly, effectively sent and received—and then merged, on the receiving end, with other bundles of the message. All of this means the message can be managed effectively for quality. With MPLS, you can throw a dozen different signals into the blender of a broadband connection and then, once they get where they're going, extract the pieces and reassemble them into a clear, clean signal. It's like being able to send a Fed Ex to a particular address, but with a note attached, "Please leave this outside the garage or inside the screen door." The shipment gets exactly where it needs to go without damage. That's how MPLS delivers an Internet phone call or message.

So T.J., and others at PAETEC, are champions of MPLS. But, in T.J.'s case, there's more to it than working with others to design and sell the best MPLS service in the industry. He thinks it's his responsibility to look ahead for the pitfalls a customer might encounter after having signed up. He's worried about quality of service, even though it isn't in his job description and will not make him a more popular guy inside the company. He has been pushing for "pro-active monitoring" for MPLS, while others, who have just as strong a passion for keeping us profitable, short-term, don't see the cost justification for it.

It's a fine mess T.J. keeps getting himself into:

> If you go in as an account team and sell them a data network, you can't tell them we aren't going to be watching the quality of the local and long distance service we provide. We do this for traditional voice service. We have a Network Operations Center (NOC) that monitors our traditional service 24/7. With MPLS, we've been answering problem calls within thirty minutes. We don't see problems and head them off before the customer does. That isn't going to cut it, even though, once we know there's a problem, we can go so deep into that customer's service—remotely from the NOC here—that we can tell them their processors are overheating, for example.

> MPLS is growing fast. We projected $800K and we sold over $2 million in 2004. In 2005, it was close to $8 million. We've already lost a couple customers because we weren't monitoring their service in real time. Here's the bottom line: we don't have a small, separate NOC to monitor MPLS. We should. My point is that we're at a critical mass for a proactive monitoring group that is a separate NOC with separate people to monitor our MPLS service. This is expensive, but it's the future, and it's integral to our brand: we are the company

that goes overboard in giving a customer that kind of personal attention. We pick up the phone after two rings. They get a human voice, not a recording. I'm one of the people pushing, on my own, to get approval to have that head count added.

I lost my cool on a conference call yesterday. I said we have no choice, we have to do this, and Sharon LaMantia was backing me up. I mean, with the kind of people we had on the line, I was sticking my neck out, getting really demonstrative about it. I called her afterward, and she said, "You know we can have all the waitresses on the floor taking orders but if you don't have cooks in the kitchen they won't get their food." I said, "Exactly. We say we're going to be a data company and it will be half our revenue by 2009, but our actions don't back it up." Sharon said, "You're right. You care. You know where we need to be to sell and support data, and you don't see those things happening."

At some point, when it becomes the best choice for everyone, we will do what T.J. thinks we should do. Until then, it's good to know he'll be banging his head against a wall by speaking up passionately in favor of it. The reality is, our service is still so good, even without a NOC dedicated solely to MPLS, we've lost only a handful of MPLS customers, and so, as much as T.J. is absolutely right about the long term, the changes will happen only when the benefit for us and our customers matches the cost of doing it. When the need is strong enough, we'll implement it.

What matters here is that T.J. is fighting for quality in an arena *where he isn't required to put up a fight*: his job is to help steer the company toward the right products for the best possible future. It isn't to worry about the quality of service we provide after those products are sold. But he does worry, because he takes the future of the company—and the customer's perception of PAETEC—to heart in a way that's much deeper than you'll find in people doing their jobs at most other companies. He behaves in this way because he knows he has joined a company that is, after only eight years, striving to give birth to itself a second time, and he wants to be a key player as the new PAETEC emerges. It's the spirit that saw us through the first five or six years and it's coming alive again now that we're at the beginning of a new era in telecom. We're engaged in remaking ourselves and our relationships with customers, so that we can become a prototype of the New Telecom: a company that shows customers how local, long-distance, and data can be done in a new, better, integrated, way. T.J. knows that he can make his mark at this company, can help steer it toward a better future, and see the consequences of his work for years to come. T.J. is writing his initials into the cement while it's still wet.

> We get to help determine the direction of the company which you can't do anywhere else. The people here are great. No matter what meeting you're in, what initiative you have, you're able to contribute as part of whatever you are trying to do. It is always like a committee: you can bring out the pros and cons. In a larger company, it's rubber stamp. People listen to each other here. You can change things. You can have a major effect on the company, no matter who you are or what you do. Even though I've had a couple offers and could make more elsewhere, I'm sticking it out here because I think I could eventually be a leader on the data side.

> Whether it's future technologies, planning on that side, or helping sales position data—I feel I can be part of the company's mission and have an impact on it. It isn't about money, or options. It's about having an impact and being recognized for it. When I'm older, and PAETEC is really successful, I can say I helped shape the company. I have my allies. Here, you can get four or five people together and, if you have this unified approach, you can even get top executives in a room and gang up on them, and they'll listen. I'm not in the driver's seat, but I'm in the back seat—people who are driving are listening—and I can have some sort of say in how the company gets built. If I didn't care, I wouldn't have frustration. But I seek out frustration, because that's how I can have an impact.

All of this demonstrates the depth of ownership our people feel. What makes it possible for T.J. to put that sense of ownership into action is his confidence that he can say anything, take a stand and voice his opinion with anyone, from me on down the organizational chart, and be respected for it. It's about open communication. He's free to express what he thinks and feels, and he knows we'll listen, even if we don't act on it. Our doors—and our ears—are always open.

This kind of unfettered communication is the root of both a sense of ownership and leadership at all levels, the sense that everyone in the company is empowered to take responsibility for the entire company's success.

One of our people in New Jersey described how I encourage candor from everyone: "Arunas doesn't want you to perfume the pig. He says, 'You make a mistake, you tell me right away. It's OK to make the mistake, just learn from it and avoid letting it happen again.' The reality is that you can make the same mistake a couple more times, usually. Leadership at PAETEC keeps the faith in its people a lot longer than at most companies." And it's absolutely true: we hold onto people, and believe in the ability of our people to get better, with far more patience than most companies demonstrate. We fire people every year, pruning the rose bush. But our people know they get third and fourth chances to get something right that they wouldn't get elsewhere, and it's absolutely essential that they do, because they wouldn't be honest, they wouldn't own up to mistakes, and help the company learn from them, if they didn't feel the kind of trust this sort of tolerance supports. They know they can be honest and say something unpopular, and still hang onto their jobs.

"PAETEC is very good at listening to employees," one employee says. "Even the VPs and managers listen to you."

We respect everyone's opinion, because attitudes and opinions are what govern motivation, commitment, and the sense of ownership we prize. I'll let any employee challenge me on anything I've done, any decision I've made. If it's important enough to that employee for him or her to have an opinion about it, then it's important enough for me to explain and explore, with that person, why things are the way they are.

Here's how one of our senior people put it:

> Arunas is willing to debate with a line manager over why renting a corporate jet occasionally makes more sense than riding coach with the general public. They'll have a spirited debate over using a charter or not, and it's our way of being open about everything and respecting everyone's opinion. Arunas will bring up something on a conference call: "Senior officers will be getting such-and-such for a bonus." People are constantly saying that, no matter how they feel about the company at the moment, if nothing else, at least Arunas is honest, at least he's telling us the truth. If I'm at a Customer Advisory Board, I'll often get questions about our profits. I tell them exactly what we make. I lay it all out: "Here's what we made last month." I don't want an obscene profit, but a good profit. We're willing to talk with anybody anytime about almost anything. Everybody wants to have conversations, so most CEOs would consider this kind of open-door policy a burden, but I've never seen Arunas reject a request for a sit-down. People want to come in and talk about their personal lives. We listen.

> Even when information gets out that we would prefer not to make available—for reasons of privacy, for example—we don't run around with our hair on fire looking for someone to punish. We have nothing to hide.

This is kind of candor is not easy to maintain. If anything in a company's culture tilts toward fear or resentment or anger, people often clam up. I don't want, in this book, to paint a picture of a Utopian work environment. Our growth hasn't been smooth and untroubled. Like all companies, we've had our rough stretches. In 2005, the mood in the company started to go south. It became clear that I had to make some changes, and, when I did, part of what we discovered was that, having gone through a very difficult period, everyone on the senior team learned how to be much more candid and direct with me. Many of them had come so close to leaving the company, during these darker months, that they established in their own minds the facts that a) they were willing to leave under certain circumstances, and b) it was safe to be completely honest with me about how they felt.

Chris Bantoft, who heads up our agent and wholesale sales channels, is relieved that PAETEC has gotten through this dark period and come out clearly committed to the values of this book, and he believes the experience has only helped strengthen open communication where it counts most, at the top.

People got to the point where they were so troubled by what was happening in the management structure that we all got much better at saying what we really think—now that we're past it. Even at the officer level, people who kept quiet before the management changes will now speak up. Everyone is thinking, "I'm not going to let this happen again. I realize what my bottom line is: under what conditions I'm not prepared to stay."

Now communication is better than ever. E.J, our COO, is a big part of this. He was at a show and he just went over to an agent and spoke to the guy for ten minutes and the agent was astonished. He'd never had a COO come up to him and talk that way. E.J. is a great communicator. A bright guy. When we had to cut costs recently, E.J. stood in front of the whole sales force in the South and said, "I am responsible for the cuts in your pay." That changed the whole mood in the group. They respected him for that. We had a regional function in the West, ten days after Arunas made crucial changes. The spirit was incredible. I said to a senior account manager, "I know you are thinking of leaving." The whole place was back to what it used to be, and it's a much healthier situation.

Arunas is being applauded by a lot of people who knew we cancelled the IPO because Arunas didn't want anyone to lose a stake in the company. This year another feeling is growing: this year's budget is an investment budget. The focus isn't on numbers that would enable us to go public successfully, but it's on providing value to customers. Arunas recognizes that we need to invest in values and service. The board is a bit uncomfortable with that. But at the senior officer level, there's a feeling that the guy's doing what is right, and we're back on track. There's a feeling again at all levels that, "This is *our* company."

T.J. Kulpa explained the sense of ownership that grows from open communication in a story about email:

> Last weekend, I was watching the Syracuse game, the kids were playing with their cousins, and my wife was reading a book. The thing is, I'm watching the game, everything's great, and I actually wanted to get up and check my email from work. I *wanted* to. It was nine o'clock at night, and I saw something come in from other people, so the other nuts were on-line too. We were talking about some problem we'd been trying to solve. "OK, we could try this." That sort of thing. You wouldn't do that if you didn't love your work. It's almost a hobby. There's a play element to it, like a video game: "Let's see if I can solve this. Let me think: I'll give him this site, have him download this, and I'll upload to the FTP. Let's see if that works." It's like a puzzle. Last night, I couldn't wait to get in here to see the results of what I'd sent the last time I was on-line. It's eagerness. It's fun. It isn't about anxiety.

Telecom is just an excuse, a pretext, to achieve what he's describing here—a culture where work can be a way of life, where people *care*, and where now and then it's as fun as playing a game. That feeling is the *real* achievement. We could be selling vacuum cleaners, motorboats, raw manganese, what we sell is beside the point. This spirit of ownership is the *real* product. This spirit is what customers are really buying. Telecom they can get anywhere. Not this kind of spirit, though—you come to PAETEC for that.

I've been talking about one side of this open communication equation: the freedom to say what you think. Here's the other side, the listening part. A couple years ago, while reading one particular job evaluation, I recognized a muted cry of frustration on page five, the self-appraisal section, of Mike Trippany's performance evaluation. In most corporations, such a sound might have been considered the white noise of success—a gratifying indicator that Trippany, a mid-level manager, was working hard outside his comfort zone. Not at PAETEC.

Trippany was nervous. It was obvious. Someone else had become director of the Network Operations Center (NOC), where Trippany worked, and he was worried that he had no way to move up. I found him in the NOC and took him into a vacant office. I thought Mike was genuinely nervous. And he was.

"You're one of the keys to future growth," I said. "You have a future here."

"It doesn't feel that way right now," he said.

"Well you do. You know how valuable you are to the team, don't you? You can move anywhere in this company."

It didn't amount to much more than that, but it was all Trippany needed. It mattered that the CEO read his appraisal and cared enough about him to find him and talk things over. Communication was the key: he wrote about how he felt, I read it, I responded, and we talked it over. I paid attention, and I cared. What I said was almost beside the point. I'd listened and responded, and we worked it out.

Another time, after one of our middle managers was promoted, I paid an informal visit to other managers at the same level and asked if they understood how much we appreciated what they were doing, even though they hadn't been chosen for the CEO promotion. My message was: "I want to make sure you're comfortable with this. I don't want you to think you're being overlooked." This isn't being soft, it's being aware of how an employee's feelings govern his or her motivation and sense of commitment. Though none of my people told me they felt overlooked, one or two have said that this sort of care and level of communication makes PAETEC different from anywhere else they've worked. Again, it's about putting people first, and communication is one of the primary ways you show how much you care about your people.

FLATTEN THE ORGANIZATION TO ENCOURAGE COMMUNICATIONS

At PAETEC, the people at the top are perfectly willing to put themselves on an equal footing with any employee in the company. We don't talk down to people. And we don't stay at home watching TV when our people are running around, trying to restore service to a customer. We aren't fire hydrant painters. When I attended our Customer Advisory Board meeting in Orange County, one of our customers came up to Bruce Peters—the consultant who originated this program and helped us run it—afterward, with a hint of tears in her eyes, and said, "I've never been treated this way by any executive, in my own company or any other company."

I like to think her immediate emotional response had something to do with the fact that I hadn't shaved, wasn't wearing socks, and my shirt still had wrinkles in it from the way it was folded on the store's shelf, all because my connecting flight in Chicago had been cancelled and my luggage still hadn't arrived. (Not having a corporate jet makes life difficult at times.) It probably had more to do with the fact that I sat there for three hours and just listened, except for the five minutes I was given to introduce myself. She knew I was not the sort of person who considered himself superior to her.

This is what I expect, and demand, of every person on the Senior Team, and every person who supervises anyone else at PAETEC. It's more than attitude, personal style, and compassion. It's about a willingness to act in any capacity if the moment calls for it. I'm going to offer a wealth of examples of this in a later chapter, where I describe how we mobilize, as a company, when a customer's service goes down—and I mean WE, all of us, from the top of the organization on down. But it doesn't take an emergency. We will do a job that needs to be done, if it needs to be done, in any circumstance. Though this is not how an organization can operate when it grows past a certain point, it is how our company has grown so fast and so successfully up to now. By flattening the organization and making everyone feel as if we're on the same level, in some fundamental way, we breed open communication and that breeds teamwork.

Not long ago, I was paired with a group of sales people in New Hampshire, and they were visiting customers. Their service from AT&T was down the morning I arrived. I was there with our sales staff, on a tour of customers in that region, to show my face, say hello, and be available for any questions they might have. The timing couldn't have been better, from my perspective. They told us, "AT&T can't even place a 'trouble ticket' live. We get an automated response, and it keeps rejecting our phone number. It's totally impersonal." That was 9 a.m. Later in the day, we checked back, and they hadn't heard anything from AT&T even by 2 p.m. We went in and I said, "We don't have that kind of system. You'll hear a personal voice when you call, and we'll take responsibility from that moment until the problem is solved." I was essentially saying, I'm the CEO and I'm telling you this. We care that much about small customers like you. What more could they ask for? "We'd never get the CEO from AT&T out here," they were saying. They signed up with us. Another CEO might not have cared enough to put on his sales hat for the morning and deal with an issue like that, but I was there, and it was the perfect time to pitch our service, so I did. I was a PAETEC worker first. Being CEO was beside the point, for me. If a customer was in need, my job was to relieve the pain.

Much of what we do to level the organization is symbolic. I've driven to New England and the East Coast—a full day's drive—many times. (If I can double up, having someone else drive, I can get work done in the car.) We still don't have reserved parking spaces. It matters to people. It's only a gesture, but it means something.

One of our marketing people puts it this way:

> We were all on a trip to Florida to some management meeting and I saw all the executives going to the back of the plane, sitting by the mothers and their babies. Arunas understands it's never about one person. It's always about the team. The team makes you or breaks you. The team does it. He's one of us. He appreciates what we do so much. I believe he does.

It's all about attitude, and one of the most important ways we express the sense that we're all in this together, as equals, on some level, is in our sense of humor. If you don't have a good sense of humor, you don't quite fit at PAETEC, because humor is a great equalizer. Especially self-effacing humor—you have to be able to laugh at yourself, just a little, here. I could give you dozens of examples of this. Once when two of our customer service managers printed out pictures of one of our executives for Halloween, glued them to masks, and then went around the entire company doing a drop-dead accurate imitation of this particular executive, complete with coffee cup and rapid-fire comments through doorways. Another time, at that party in a local hotel, our gondola time trials got us kicked out and permanently banned from the premises. Or the time when one of our people in Collections said, "If I get this customer to actually pay his bill, I will do a victory dance in the NOC." Everyone in the NOC was laughing when he stood on a chair and boogied.

A female attorney who works for us says:

> We had a conference call not long ago during an extensive negotiation. It was with a white-shoe law firm in New York City. Al Chesonis and I were on the line at this end. At the other end, these guys were taking themselves very seriously. So I figured it was time to put them on a little. Break the tension. You know Al: he's the nicest, most sentimental guy you'll ever meet, right? I knew one of the women at the other end had met Al. So I said, "I'm going to be willing to give up on some of these issues because if I don't, Al is going to hit me." There was this awkward silence. Al pipes in: "I'm a rather large man." At this point, the woman jumps in: "Oh, I've seen Al, and you should be afraid." The three of us were laughing hard, and the others didn't quite know how to step out of that. But it was disarming. We came to an agreement quickly after that. Where else would someone in my position be able to talk that way about a senior vice president?

When you have top executives willing to do a job that would be beneath consideration for most executives at most corporations, and willing to be the butt of jokes from nearly anyone else in the company, it inspires workers to behave as if *they* were CEOs. That kind of tolerance for humor is a way of communicating an underlying friendship that's essential to teamwork.

LISTEN AND THEN ACT ON WHAT YOU HEAR

Open communication is formalized in a number of ways here. Regularly, we send out anonymous employee surveys to get as accurate an idea as possible about how our people like working at PAETEC. It matters enough to senior management that we've spent a significant amount of time urging people to respond to the survey. It's a very successful program with a response rate usually around 70 percent.

To get completely candid responses, we utilize outside resources to gather the survey results and report back, but we analyze the findings internally, after all details that might identify the participants have been removed. We ask people if they are satisfied with their jobs, if they feel welcome when they go to a manager with an issue, if they are confident in PAETEC leadership, and if their benefits are satisfactory. We ask about larger issues as well: Is PAETEC really focused on customers? Are we providing people with the knowledge they need to do their jobs? And would people recommend PAETEC to someone looking for employment?

Once we have the findings, the senior team meets to talk about possible courses of action. We don't just acknowledge issues raised in the survey. We take action. When our people said they wanted changes in benefits, we altered the plan. We have a self-funded plan for medical and health, which gives us leeway in how we structure benefits. When people said they wanted increased coverage for vision, we offered better coverage for eyeglasses and contacts. When they said they wanted better dental benefits, we gave people more help with dental care.

Before we take this kind of action, though, we respond to the survey: we turn it into a conversation with our people. We put together PowerPoint presentations on the survey results and post them on our internal website, where we respond to many of the most common comments. I will address the survey results in a company-wide "Town Hall Meeting," where I describe how people responded to the survey and what we hope to do in response to their concerns. I also talk about the surveys on our regular Friday conference calls and management-only calls for the approximately 150 people on our management team.

CHAPTER 5
Tell More Than Most

T HESE CALLS ARE A PRIME EXAMPLE of how we try to communicate everything important to everyone, inside and outside the company, all the time. Top leadership sets the example by being open, listening, and speaking honestly.

When I conduct our regular Friday company-wide conference call, I stand at a podium, surrounded by any employee who wants to be there in person, in the atrium of our headquarters. It's a two-story space with a mezzanine looking down on the entry, where I stand at a lectern beside the front door. Behind me is a gently curved wall made entirely of glass that looks out onto the small valley behind us with hills on the far side dotted with houses, a nine-hole golf course, and office space—the office space where we began. Several hundred people gather on the ground floor, sitting on the curved staircase, leaning against the mezzanine railing, listening as we send my voice, or the voice of anyone who steps up to the podium, out over the network. Anyone in the company can dial in and listen, and can also comment, or ask questions. It's informal, and I always end it with a joke.

Easily, a third to a half of our employee base participates in these Friday conference calls. I talk about anything: financing, acquisition plans, profits, bonuses, stock options, and I recognize people for their contributions. In fact, these Friday assemblies are a prime way we offer recognition: anyone is welcome to get up and honor anyone else for the work they've done, and when you have most of the company listening to what's said, it can mean a lot to the one being recognized. Much of what I tell our employees, as well as anyone else who wants to dial in—customers, vendors, anyone we invite—is usually kept in the board room at most companies. This policy is a show of respect for all the people who make PAETEC great, and it's also a way of setting a tone of honesty: we have no dark secrets here, and even when the truth hurts, we tell it. We don't want our people to be taken by surprise.

Even so, the concluding jokes are often what people most want to hear. John Budney used to provide most of them. After he'd discovered he had cancer and was unable to make a personal appearance he would dial in. Once, during that period, he even joked about his health. I asked him, over the speaker phone, why he hadn't shown up for the call. "I can't make it today. My doctor says I have myopic proctologia." I waited for him to elaborate but he didn't. "I see. What is that, exactly?" He said, "It means I can't see getting my ass into work."

I am as open with our customers, personally, as I am with our employees. Every quarter I send a brief email out to all our customer contacts. I try to keep it informal but highly useful. Possibly the most significant information in each of these emails is simply my email address and phone number: it's my way of reminding all of our customers that I'm here. I answer my email and I pick up my phone when it rings. My excuse for reminding people they have this kind of personal access to me is an update about a service or product, or, on occasion, simply contact information for someone else at PAETEC, such as our vice president of billing services. (When I initially sent out this individual's contact information and invited customers to call him directly with questions, the response was enthusiastic.) The last line of the email is always the best: "Please contact me directly at 585.340.2567 or arunas@paetec.com if you have any suggestions for our company."

Being completely open with everyone about everything isn't possible. Keith Wilson, our CFO, describes how hard it can be to tread the line between what people need to know and what they can't, legally, be told:

> The more people know and the better they understand, the better their business processes can be. But obviously, we're a little more cautious about what Arunas should communicate: he's an open book. One of the positives about this is that Arunas is passionate about being extremely forthcoming to the employees. You can imagine why this might make his attorney and his controller a little nervous, so we try to balance his candor with caution. It's a tough balance using open communication because a lot of the things I deal with are confidential. I work with Dan Venuti, our general counsel, to help Arunas understand what he can and can't share.

And yet we are as open as we can be within the law, and we've tried to open up lines of communication in ways that would never occur to many companies, between, for example, our auditors and our employees. We give all our employees access to our auditors every year, and they can ask any question about finances that they want. They can send questions ahead of time. The senior officers stay off the call, so as not to inhibit anyone. The first time we offered this forum to employees, we had a hundred people listening and asking questions. A couple years ago, it was down to twelve. At the start, with all of the scandals in telecom, they were nervous about our finances. Now they know they can trust the company they serve. As Keith Wilson puts it:

> Every year employees have an opportunity to call in and ask our auditors whatever questions they want. In an industry beset with financial and accounting irregularities over the past few years, this builds enormous trust. There's only so far you can open the kimono, though, under SEC reporting regulations. That has prevented us from being as forthcoming to the public as we would like. That's tough because there is such a culture of openness.

WE'RE AS OPEN AS WE CAN STAND TO BE

In all of these ways, the most important quality we try to uphold is being honest and open—making ourselves accessible, to one another and to the customer. But part of what's different about this openness at PAETEC is how it works its way into the emotional tenor of our conversations with one another. It's grounded in our caring culture. Those of us who "get it," those of us who understand the level of caring here—about our mission as a company and about one another as people—understand that we can have fierce conversations and be blunt without much stroking beforehand. It means, above all, being candid and direct, to the point of annoyance on occasion, but with a deep sense of commitment and caring beneath that honesty. It means saying what you really think to others, but also listening with tolerance when someone else is being blunt with *you*. It means everyone's door is always open.

This means listening to things you might prefer not to hear. "PAETEC has a very open policy. No senior manager, director, or VP gives off the air that he or she is untouchable," says one of our West Coast directors. "People's doors are always open. I don't have to be afraid to speak up. Our opinions matter. You wouldn't see that in the majority of companies."

With your family and your best friends, you can kid around and be serious at the same time. That's the tone we encourage. Jill DiVincenzo, our benefits manager, associates this, of course, with her days working for John Budney:

On my first day at PAETEC, I met with Arunas, whose newest baby was hanging out with him in his office for a while, and he paid more attention to her than he did his laptop. I knew I'd made the right decision. Not long after that, Arunas threatened to fire me if I remained the receptionist much longer. While his frequent threats made several eyebrows go up of those around who heard, I knew he saw potential in me. It was a compliment.

On my first day working with John Budney he grinned and told me I was an ass! That is "Agent Support Specialist." I never laughed so much and learned so much in such a short time as I did working for him. I think I can still feel him pounding the back of my chair and shouting, "I need that report now!" as if the pounding (and yelling) helped. On one Bosses Day when I gave him a card and wrote in it, "You are a pleasure to work with, John," he called me into his office and with card in hand hollered at me, "You work FOR me, not WITH me!" He laughed and laughed. I have yet to meet another like him.

Then Jolanda Chesonis set her sights on expanding the HR department and wished for me to accept a position working with her—I mean, FOR her. I accepted and here I am today, still loving it, still knowing I am helping people and never, ever seeing a "closed door policy" of any level in this organization.

The day I was chosen as a John V. Budney Award recipient, for the spirit I bring to my work, I cried. I knew I was appreciated and valued and still know it today. I also know very well every one of our "higher ups." I respect them for their candidness, kindness and welcoming demeanor. Their doors are never closed.

THE BUDNEY FACTOR

Every family needs a big brother who leads the way for everyone else. When it came to open communication, John Budney served that role for us all the years he worked at PAETEC. He wasn't exactly a role model for everyone, but he set the tone that somehow made it OK for people to have any kind of conflict and still be family. His behavior imprinted itself on all of us who had the privilege of working either with him or for him. One VP remembers the perfect example of this: "Once, I was on the phone in sort of a screaming match with MCI because they'd put one of our customers out of service. John was sitting there in the office, and he asked me to mute the phone. 'You want to see the most beautiful kids in the world?' He got out his wallet and showed me pictures of his kids. Everybody loved him."

He made it possible for everyone to have "fierce conversations." He initiated about half of them himself. One of our vice presidents in Irvine remembers an example of how he delivered a scathing assessment of what I, and a group of vice presidents, were doing to our agent sales force:

We had twenty vice presidents in sales. Four of them were from the agent sales channel, and sixteen were from the direct channel. So the agents had no sway in a vote. We're in a hotel room, at a long table, pitchers of water, bagels. Budney is sitting against the wall, hung over, his arm folded over his head, like a duck sleeping with his head under his wing, recovering. The discussion proceeds and it comes to a vote on whether they should add the financial vertical market to the others: education, government, hospitality. The vote is clearly in favor of doing so. Vertical markets, as a whole, were off limits to agents, except with permission. It was just another way of making agents feel like second class citizens, even though agents often had much closer—albeit gray-area—relationships with customers, because they often represented other vendors, in addition to PAETEC. Agents, though, were street fighters, and they knew how to close deals, one way or another. So we went around: Aye. Aye. Nay.

Arunas says, "It seems we have a majority for including financial. John, what are you thinking? We haven't heard from you."

Budney suddenly whips his arm and hand away from his head, snaps his head around and begins screaming at the group: "These VPs don't know what they are doing. Direct doesn't know how to sell. They just want to cover their asses because they can't sell crap! And, once again, they're finding a new way to stick it to the agents. That's what I think." He curls himself up, resumes the immobile position he had before, almost a fetal position, and hides his head against the wall. Everybody had to laugh, but he also got his point across—expressing a feeling a lot of people shared, which was the key here—and he was genuinely angry. But he did it in such a theatrical way, it didn't destroy the mood. It made it acceptable. Of course, it didn't sway the vote either.

It didn't sway the vote, but it opened up the communication. It got things out in the open: the anger, the resentment, the conflict. And, once that happened, we could move on. One of our VPs in New York City remembers:

PAETEC is a family, and you can have fights. I'm very loud. Budney was in Rochester and I'm in New York and we would be screaming; people must think we're killing each other. He can scream at me and hang up and then we go have a beer. If we were at one of our large competitors, he'd probably get fired for that. It's like a fraternity. There's a lot of people who put a lot of hours in here because they believe in what they do. There aren't many employees here who consider this just a job.

John knew that, to preserve the sense of family, you had to find a way to make it OK to be angry and have loud arguments. If you don't, the anger goes underground and erupts in more destructive ways—people quit, productivity suffers, quality goes out the window. If you don't find a way of making it safe to be angry, to voice the truth, the way it can be in the best families, then letting people vent does no good either. You end up with the same lousy results. The key was humor. John made people laugh.

My sister Jolanda recalls, when she joined the company:

In those times everyone was sharing offices, and my office was right next to John Budney's, and I thought, who is this individual? I'm talking to newspaper editors on the phone, and I'm hearing this very boisterous person on the other side of the wall. He had a great sense of humor. He could really play you. You didn't know if he was joking or not. First two weeks, he'd say something, and I'd think he was joking, and he'd make this face, and you'd know it was a joke. It was funny because John was very vocal, and probably a month or two after I'd been there he was having an argument on the phone and he was really *really* loud. I was used to him. He came into my office, and said, "Jolanda, if I was a little overboard, I'm sorry." I said, "Oh, were you yelling? It didn't even faze me, John." His intensity was so typical of PAETEC. Intensely committed to customers, service, and employees.

CHAPTER 6
Share the Knowledge

Iᴛ'ꜱ ᴍɪᴅ-ᴅᴇᴄᴇᴍʙᴇʀ. Tracy Robertson arrives at PAETEC around 8 a.m., with her daughter, Taylor, a freckled six-year-old. They're both fighting colds, so Taylor can't go to school, and there is no one to watch her on this particular day. Family comes first here, so Tracy is welcome to bring her daughter in an emergency. Tracy clicks through maybe a hundred emails, ignoring most of them, and lays out two Flintstones vitamins, both of them for herself, because Taylor has already had her vitamin. Tracy drinks a caramel macchiato from Starbucks, chews one of the vitamins, meets with two or three of her people briefly for updates on their progress, chews another vitamin—the coffee and the Flintstones will be her only nourishment until dinner—and applies hand lotion as she engages in a conference call with marketing staff in other locations. By 8:45 a.m. she's ticked five or six items off her list. All of them are about making communication with customers more and more open and productive at PAETEC.

At nine, she meets with Jack Baron, her boss and our Executive VP in charge of marketing. In the course of the half-hour meeting, Tracy says: "In the branding work we've done, what concerns me is that customers care about how they're treated. I think we might lose focus on that in the way we isolate what our brand means."

"It's all about our people," Jack says. "Customers will tell you we've got great people."

"Are we de-emphasizing that?" Tracy asks.

"Why do people really buy from us? When you talk about personalized solutions, we sit down and arrive at a solution that is ultimately better for the customer. We're not there to push a product, we're there to consult. Maybe that's what every great company does. But it's the person who's doing this that makes the difference, isn't it?" Jack says.

"Let's say Company X and Company Y both have good service and a good price. Who do you buy from?" Tracy asks.

"The person you like. The person you believe in the most."

"Yes. It's the *person* you pay for," Tracy says.

"But it's true. I think people would say we provide exceptionally strong and personal service which is about being a trusted partner. A trusted friend delivers on promises. Personalized solutions tie in there as well."

In other words: we make business *personal*. To recognize and stay focused on this, and to constantly try to see how work can somehow come around to expressing this, is part of Tracy's—and Jack's—leadership.

Next, trying to keep her daughter occupied between phone calls, Tracy dials ten of the customers who have said they will probably attend a Customer Advisory Board meeting we will be hosting in Orange County in January. She has been doing this for days. She calls, spends five minutes charming someone thousands of miles away, and then adds that person to the list of people she hopes will *really* show up. This is not her job, but she can relax only if she feels she and her people have double-teamed everyone who might come to the board. She wants to do every last thing possible to make sure we'll be meeting with a dozen customers—not three—when we fly out to listen to them. These crucial meetings are mostly about letting customers tell us how to solve our problems, but whenever we interface with customers, especially at these meetings, we want to communicate to them how we adhere to our four central principles. It took us nearly a year to even understand how these principles—the tenets of this book—undergird our brand.

Jack Baron has worked with Gregg Lederman, founder of BrandIntegrity, a consultancy, to define and implement a clearly-focused brand identity. Gregg is helping us transform the way we understand and work in harmony with the meaning of our brand. Gregg's work has enabled us to understand which key elements of our culture distinguish us from our competitors. Through a series of workshops, we've refined our understanding of what makes us who we are, and, over the next few years, we'll analyze our work processes to see whether everything we do is consistent with these key elements: if we're actually *living* the image we and our customers have of us. Out of those workshops came an understanding of the four sections of this book, our key brand elements: caring culture, open communication, unmatched service and support, and personalized solutions.

Open communication, the second element, is what Tracey's entire role at PAETEC is about. And her efforts in this one instance, as minor as they may seem—or extreme, if you look at it another way—are crucial. If she doesn't make these calls, if she doesn't sit down with Jack and refine how we present ourselves to customers, then our communication with customers won't be effective. But thanks to her efforts, as it turned out, we had a powerful session and a great turnout in California, and more than a dozen customers went away feeling even more loyal to us than before the meeting. They communicated who they were, and we did the same, and our bond has grown stronger as a result.

Communication at PAETEC means many things. We openly share information with employees about our company, and its successes and challenges—not only during the Friday calls, but in email newsletters, in our constantly updated internal website, and when I speak at almost any gathering of workers or customers. Most importantly, we provide our customers and partners with access to our knowledge and expertise, as we listen to their needs. This is nowhere more effective than at our Customer Advisory Board meetings, which Jack and Tracy handed off to Brian Benjamin, not long ago. He is helping to expand and improve this formalized way of listening closely to customers—letting them help guide us and our partnering companies toward more effective ways to serve them.

We have a series of webinars, seminars delivered over the Web, for all of our sales channels: wholesale, agent, and direct. We run a different one every month, highlighting a different product. Invitations are sent out ahead of time, announcing the subject, and participants register via email. A live speaker gives the seminar, and participants watch on their computers, via web cast. Likewise, customers can sign up to see a product demonstration and learn more about various products.

Again, this is where one principle dovetails with many others in this book. By sharing information, by being open in this way, we stimulate a feeling of ownership among our people and inspire leadership from all levels. People who are informed and who have access to the information they need have the courage and motivation to make decisions and are more willing to be accountable for them.

JOB SHADOWING

We encourage everyone at PAETEC to learn as much as possible and communicate as much as possible with other departments in the company. This can be as extreme as working temporarily in another department for a brief—or long—stint or simply bringing in someone from another department to speak to a small group over the lunch hour.

The way Marion Wyand, our vice president of engineering, trains her people in engineering to understand the role played by other key departments is a great example.

> I've had all my teams sit with Order Processing because I told them I want them to know what's involved in it and to understand it isn't simply an order-entry role. The people who process orders get paperwork that's a mess, sometimes. They have to figure out what the sales person is saying. What the customer really wants. What the network can actually do. In the notes of an order it might be obvious we've sold something we can't do. So we spent time with Order Processing people.

> We did the same with the NOC. At one point, we got feedback from the NOC on the systems engineering survey, and it said some of our engineers weren't always as available as others. I asked the NOC why they felt we needed to improve our relationship with them and they gave me some examples. So I had those engineers spend a day with their technicians and see what it felt like. All of this increases the flow of communication between departments and that means better service for the customer.

An extreme example of this kind of inter-departmental flow of information and training comes from one of our managers responsible for almost anything that helps our customers and our sales force understand and connect with one another. During the months of ramp-up for our new customer database, she spent two months, full time, working with the IT department, learning everything she could about how the database would work. For much of those two months, her staff were left to manage themselves because she was completely dedicated to the temporary assignment. It was a testimony to the expertise and dedication of her people that they were able to work together and make their targets while she was away, since she completely immersed herself in her new work. (There's certainly an irony in that, since we're talking about communication, but the lines of communication *within* her team sustained it during her absence.) Her productive "sabbatical" made a huge difference in her ability, now, to fully exploit the capabilities of our database as a means of better communicating with customers.

THE DANCE CARD

Often what counts most are the *little* ways you follow the principles in this book. The imperative to communicate, face-to-face, on a *personal* level, with customers extends to parties, events, and even minor social occasions. We conduct various events, and invite customers to them, around the country, all year long. Before an event, one of our marketing people will take time to make up a list for me—or other senior team members—of customers to speak with during the event. For a certain size customer or potential customer, it's a great chance to talk with one of us and get a feel for what the PAETEC leadership is like. We make up what I call a "dance card" beforehand. Once I have a clear notion of who is going to attend, she'll ask which customers could use some personal attention. Who has an issue about service? Who is ready for an up-sell? Often, people will RSVP at the last minute before the event, so my deck of cards will get shuffled.

At most companies, would senior management make an effort to talk with all twelve people listed on a card made up by someone in the sales department? At PAETEC, I do, and so do all the others on our senior team. We make sure we know the people by name when we introduce ourselves. We memorize the central issues that need to be discussed. If possible, we learn, beforehand, anything we can about their interests, their family, their personal situation. It's natural and, honestly, it's easy. It's essentially what any friend would do, without thinking twice about it, and long-term friendship is basically the outcome of putting all the principles of this book to work. It isn't just business, it's personal.

Usually, I'll get into a conversation with someone for ten minutes and someone will have to tap me on the arm and pull me away: "I'm sorry to interrupt but I have someone I'd like to introduce you to." I'll be directed to another customer, and before I shake her hand, I'll know that she had a gripe about loss of service, or was especially pleased with a solution we devised for her company and wants to talk about the next step with us. And it will go from there. It's all about the deck of flash cards beforehand. It's an old-school kind of database, but just having shuffled through them on the plane earlier in the day will have refreshed my memory on how someone went to school in Delaware, or just had a baby, or is a scratch golfer. Our sales reps and our senior account managers have a deep knowledge of all our customers, so we're able to rely on what we know about our customers and communicate with them even more effectively than before.

KNOW WHO YOU'RE TALKING TO

Until now, the knowledge we accumulated about our customers was spread throughout the company in various ways: on individual computers, in paper files, or in the heads of our sales force. We decided we need one central repository of all information about our customers, a database almost anyone can access when dealing with a customer. We knew that just having the simplest information in one accessible place would improve everything we do for customers. We aren't talking about a customer's birthday and favorite color, but name, address, email, what products or services that customer buys from us, and any other information that could be useful to us as we serve an account.

When we bring in a new customer, he or she registers with us and elects to receive various kinds of communication: CEO mail, webinars, PAETEC Online access. Six months after we cut the service over to PAETEC, the customer gets a customer satisfaction survey. A year after that, he or she gets another satisfaction survey. Have you been contacted by an account manager? Without a central database, it's nearly impossible to know what kind of information to send out to our customers, and we can't be proactive in our marketing. With the new database, we can send customers exactly the kind of information they need about new products and services, customizing how we communicate with everyone.

With the database we can also track the customers we lose and organize ways to follow-up with them on why they've left PAETEC, to help improve our service to the customers who stick with us. It enables us to automate our communication to potential new customers as well, from target lists our reps submit. In some ways, this all may sound impersonal, but in reality, the new software enables us to communicate individually with one customer at a time. We have knowledge about that individual customer at our fingertips whenever we call on that customer—or the customer calls on us.

When a problem becomes a genuine emergency, for example when you have a service outage with a real impact on your organization, the communication gets even more personal. When we've screwed up, or one of our partners has done something to degrade service, invariably we try to take full ownership of the problem, even if we're not at fault, and respond with a real human voice to the first phone call for help. From then on, we deploy as many people as are needed to resolve the issue, from all levels of management, up to the CEO. But the most important step is the first one: *we listen.*

During the worst, and longest, network outage in PAETEC's history, in 2004, customers in New England went without service for more than 24 hours. Our COO and a dozen members of our senior team were on the phone early in the crisis, talking directly to customers. The outage nearly put an ambulance company out of business, and a Harley-Davidson dealership was particularly incensed. One of our senior people talked with all of them, including some of our smallest customers, who wouldn't even register on the radar at any other telecom. Because of how we handled it, they stuck with us.

The reality is that communicating with customers this way doesn't alter the fact that we failed to provide service for an extended period of time, but it makes a difference. It means something that a top executive is willing to talk personally with an angry customer, take the heat, not make excuses, and simply apologize and pledge to be completely honest about what happened. The other reality is that in every business like ours, all providers make the same mistakes at one time or another, and what distinguishes PAETEC from all the others is how we handle ourselves during the bad times.

All of this requires us to open ourselves as a company and encourage free communication among employees and between employees and customers. The extent to which we make information available to our people, and to the public, can be almost frightening to a new employee who isn't used to knowing so much. Knowledge is power. But it's also responsibility. When you give people that much access to how our business operates, it begins to dawn on them that you expect them to take responsibility for the success of the entire organization.

This lack of boundaries extends up and down the organizational chart. One new hire, in his first week in the NOC, needed help with a customer problem: "I'd been here just over a week, and I could call a vice president and ask him for help. At my former employer we had to prove we needed to engage somebody from another department. Here I just pick up the phone. It's as if I'm talking to somebody on my level in operations."

One of our best sales reps put it this way:

> PAETEC shares information. They give us enough knowledge about the company to keep us interested. People are freaked out by how much they know when they first come here. Somebody comes here, and he asks "What price do I charge for that?" The answer is, "I can't tell you. That's up to you." This isn't what you normally hear in a sales organization. You ask that person, "What do you want to charge?" or say, "Go to the profit model and figure out what works for us and for the customer. If your customer wants two cents a minute, tell him how much business he has to give us to get to that rate." Customers are taken aback when we tell them what we pay Verizon in order to use their loop, for example. We're just totally open about all the costs associated with a deal. It makes it very clear to the rep and to the customer how much we're going to make, how much it's going to cost the customer, everything. Ultimately, this kind of knowledge creates a sense of ownership. And it builds trust with the customer. A rep

comes to work thinking he or she has a business to run, not just a target to hit. It's just weird that PAETEC tells me how we get to the final number we quote to a customer, and the customer is as welcome to that information as the rep.

LEARNING TOGETHER TO PROMOTE TEAMWORK

The more you grow, the harder it gets to keep the family together. A lot of it has to do with the way people interact, especially in the bearing of top executives toward others in a company. You can spend half an hour and tell everything about a company's culture by simply watching how a senior vice president speaks to someone further down the organization chart. This has to do with the idea of leveling the organization, and getting people to work *with* one another rather than *for* one another, but it's really about tone.

A caring culture opens up communication. One leads to another. If you care enough, you communicate what matters without hesitation. When John Budney was in the hospital, two of our people, Marion Wyand and Rita Scotto, suggested we record a video for John. I thought it was a terrific idea. I told everyone involved to drop everything and do it—this was toward the end—and I wanted the video to be delivered to John that same day in the hospital. We needed to do something special for John. Marion sent an email to all the different departments explaining where and when to meet. She included the lyrics to *Can't Smile Without You* in the email and assigned different parts to different departments. We assembled a large group of people and we sang it, while our IT department scrambled and did all the recording. After he got it, John sent us an email. "I have never felt so loved in all of my life." In the video we had said "We hope your wife is spoiling you." When he saw it, he said, "Yeah, right." He still had his sense of humor at the end.

Rita likes to change the lyrics of songs, taking a tune and making up new words that describe someone in the office. So Marion once pulled me aside and said, "Tomorrow is Rita's birthday. Would you stop by and ask Rita to sing you a song? Around ten o'clock?" I said I'd be happy to do it.

So the next morning, I went down and stood beside Rita's desk and said, "It's your birthday. I want you to sing a song for me." She was hyperventilating a little, or at least pretending to be out of breath as a way of making me go away. She kept shaking her head. She wouldn't do it.

I said, "I'll go in another cube, so you won't see me."

"I can't do it. I can't do it!" she said.

"Rita, you won't get a bonus if you don't sing for me," I said.

It didn't work. When it comes to the most crucial moments on the job, bonuses just don't motivate. This is never more apparent than when you are trying to get a woman to sing for you. So, later, I came back, after lunch.

"Rita, will you sing a song for me?"

"No."

"You are a trash talker, Rita. You talk big, but when it comes time to follow through, you just won't produce. Your bonus is on the train right now, and it's pulling out of the station."

All of this was two weeks before we announced the amount of the bonus. Rita knew we were doing a second bonus, in the middle of the year. I'd sent out an email on it, and she had taken my group email to everyone in the department, and she had replied, asking, "How is my thirty percent bonus looking?" So that's why the idea of making Rita sing me a song seemed like a great way to bond with one of our employees. She never did sing for me, but she got her bonus. But it wasn't thirty percent.

The point here, of course, has nothing to do with the bonus, or with the willingness to sing, or with pseudo-bargains I might make with somebody to amuse myself and everyone within earshot. The point is that Rita felt we, at PAETEC, were enough of a family that she could take an email I sent to others in the company, and hit REPLY, and shoot off a smart-ass comeback to the guy who runs the company. And that he would laugh and decide it was an invitation to spend a little time putting her on. That's a big part of what it takes to keep the spirit of family alive in a company.

Obviously you also have to start making efforts in a more organized way. Much of the success of this effort depends on what individual departments do for their people. Jim Manetta, who runs our NOC, is a shining example of how to build a team that functions like a family. One year he assembled his team, off site, for an entire day of group challenges, a sort of obstacle course of exercises that called on his people to work together to come up with new ways of thinking about how they all depended on one another.

He and a few others on his team put in forty hours planning for this one day.

They had a few volunteer firemen in the NOC, so they started the day at a firehall. Next, Manetta handed everyone an instruction card, with the challenge: "Go to bowling alley, and get three marks in a row." Three strikes, two strikes and a spare, whatever. The trick here was the ambiguity of the instruction. All they had to do was line up and roll three balls and get any result at all. Some tried for three strikes, but that wasn't the point. Jim just let them figure out what to do on their own: they had to define the rules themselves, to some degree.

Again, it was fun, but it was also a lesson in how to deal with challenges. The rest of the day went the same way: impossible-sounding challenges, of the sort people face at work all the time, and an imperative to work with others to figure out how to get around them. A scavenger hunt at the mall. A math problem disguised as a bucket brigade in a driveway. An impromptu instruction to build the chassis of a wagon. It was all about communication and sharing knowledge. And laughing while you did it.

"The whole thing started at noon. We got back at three, three-thirty. We had a party afterward. We cooked pasta in the big kitchen at the firehouse and had pop and desserts," Jim says. "This was our second year with this. We didn't want to turn it into a program for other departments because it isn't something that would necessarily work for other groups in the company. But if we modified it, we might be able to use something like it for the company."

The challenge is not to lose the informality—the feeling of how we work together as a family—as you grow. Peggy Stewart remembers an experience she had on her first day at PAETEC, and how she relived it recently:

I remember that first morning like it was yesterday. The place was so small and completely packed. There was nowhere for me to sit and no computer for me to work on. I ended up in someone's office that was on maternity leave. My new boss showed me what he needed me to do and left me alone in that office to get it done. I was so nervous that I don't think I heard one word he said. I felt like I was ready to cry. Then one of the provisioning girls (Robin Jones) came to check on me. I had a ton of questions, which she answered and got me off and running with my assignment.

Just as she was about to leave, someone came into the office carrying a box of candy and said to both of us, "Go ahead, take one." It was one of those boxes that they use to sell candy for fundraisers. He was handing them out to everyone. I thought, "Wow. Nice guy. Strange, but nice." Then he looked at me as if he realized that it was my first day and put the candy down. He said, "Hi, I am Arunas, I don't believe we've met yet." He extended his arm to shake my hand, and then he said, "See you around," and was gone. I looked at Robin and said, "Who was that?" She simply replied, "That was Arunas—the CEO." I must have looked shocked because she just started laughing and said, "Oh he does that stuff all the time."

Six years later, at my own desk, using my own computer, I have that same feeling of excitement that I had the very first day. I am not saying that at times it isn't stressful and that some days aren't crazy, because they are. I just think that it is because as much as this place changes every day, there are so many things that stay the same.

A couple of months ago I was sitting here at my desk surrounded by bills, and Arunas showed up with a box of candy. He said, "Go ahead, take one."

In a more practical vein, knowledge-sharing promotes teamwork on sales calls. Our reps will often bring an engineer along on a call to explain, in technical terms a customer will often respect and understand, the capabilities of our networks, our services and our newer products. As the sales cycle advances to a crucial stage, and a customer's trust needs to be nudged to the highest possible level—given some of the malpractice in the telecom industry which has gotten so much media coverage—we'll send a finance person along. Tim Bancroft, our treasurer, describes how this works, both as a way of sharing knowledge with the customer and building teamwork inside the organization:

Sales people often ask finance people to help them out on sales calls. Since we are a private company—and in some regions most people have never heard of us—potential customers don't have public access to our financial statements and information. But customers can sign a non-disclosure agreement with us in order to see that information. Then they often have questions about it. That's when sales comes to finance for help in explaining our numbers to customers. If we do a good job at this, it can help close the deal. One of the biggest users of finance in this way was John Budney. John used to joke that finance people were "his closers" because the deal usually was won right after they talked to his customers. In reality, customers are usually pretty well sold on PAETEC by the time finance gets involved, so talking with finance was the last bit of reassurance a customer needed. But the way finance and sales work hand-in-hand at PAETEC shows how much we communicate with one another and with the customer, and how it's essential to team-building.

The most important aspect of open communication is to make everyone feel as if they can know almost anything they need to know about the organization. And that means a sense that the people running the company will tell them the truth and will be keeping them posted on the latest news. Dick Padulo, our executive vice president in charge of operations travels around the country to talk with his people candidly about what he sees happening here, as do all of our senior team members. He believes open communication is at the heart of what makes PAETEC successful.

Open communication is key to a successful organization. I believe it's true that only half the herd gets the word. It doesn't matter how big our organization gets, if it's not written anyplace or covered during Arunas' conference calls, it doesn't get around.

So I try to get out to all of our switch sites at least once a year, so I can talk to people and have them ask me questions as a senior team member. "What do you think of the company?" They tell me a lot of things they may not say to their boss. They are honest and truthful with me. It's important to them to hear what the top executives think. For example, we're moving into the VoIP world, and that's a big change. The message must get to the people why we, the senior team, are making that change so that they are behind us, instead of questioning us. But we encourage people to question us, as well. I have no problem having anyone confront me about anything. If my door is closed they can just march right in, and holler at me. I believe wholeheartedly in communication, and being down to earth when you talk to people one on one. We're not two-faced. What we do is what we say, and what we say is what we do. I think that's the mentality from Arunas on down, that's the culture.

PART THREE:
UNMATCHED SERVICE AND SUPPORT

CHAPTER 7
Make It Personal

I$_{T'S}$ $_{A}$ $_{MORNING}$ $_{FOR}$ $_{EMERGENCIES.}$ Trish Dow, the senior director of customer service—one of the most unflappable human beings you will ever meet—already has fifty voicemails from customers the minute she walks in. Networks are down in various places on the East Coast, and before this day is over, there will have been a total of eight different outages. Customers are beginning to realize their lines aren't working, so the calls are coming in—and being routed to the right people.

A customer service rep takes a typical call: "This is Darlene."

"Darlene, this is Debbie."

"Hi, Debbie. How can I help you?"

"I'm calling for the person who manages our long distance," Debbie says. "She said for me to go ahead and call. We have, since Friday, maybe longer than that, been unable to send long distance faxes. Now, it's grown as a problem. In the last five minutes I've been able to send one to our New Jersey office. But I got at least four messages saying account code not valid."

"We can look at the line."

"Okeydokey. We had a major outage before the Fourth of July. I'm going to say it was on June 29. We've been having problems since then and our manager wanted your company to check all four of our T1 lines. It was only an hour and a half but all sorts of things have resulted from it."

Darlene says, "We'll issue a trouble ticket. They'll take a look at the failures this morning and find out why those lines aren't working."

"I was able to send the document finally, but it was frustrating."

"Absolutely, it is, when you go there to do one simple task and have to repeat the process so many times," Darlene says. "We'll look at the problems in that time frame and look at what's occurring and see the connection between that and what you're experiencing."

"Would you email me on that?"

"Absolutely."

All of this is happening in the background, for Trish. She has a set of meetings this morning, and doesn't expect to change any of them, even though, by the end of the day, she anticipates hundreds of phone calls from customers: problems, queries, the need for some kind of information or response. Her team of 29 people are trained to pick up the phone immediately, be friendly, sympathize, and, most of all—get our engineers, or someone else, working on the problem.

Generally, here's what happens. Her people open a "trouble ticket," pass it along to the Network Operations Center (NOC), which springs into action, helping diagnose the problem and if possible, calling back customers as often as they want—usually every half hour, but more or less often, if requested.

It's important to know that in most situations we don't own the customer equipment that causes the problem. PAETEC has managed to be so successful partly because we chose, at the start, not to own the fiber networks we use. We didn't lay any fiber. We don't own cable. We pay other companies to run our traffic over their lines. But we do own the switching network, the "intelligence" of the network's operation, and the "last mile," the T1 lines that link our customers to the larger networks. We, in turn, charge other companies for their traffic across those T1s. The balance between the savings we enjoyed by not investing a lot of capital in networks, and the revenue we brought in with phone and Internet traffic over our T1s has been a winning combination. But this is a little tricky when it comes to problems, because when a network goes down, and service to our customers fails, it usually isn't our fault. That would sound like a good thing, but here's the catch—it's our *responsibility* to help our customers manage the crisis it causes for them. We have to take ownership of everything we do for our customers, even when we aren't creating the problem. If it has an impact on our customers, we take responsibility for it.

It's the way we take ownership that distinguishes PAETEC. We will assemble the various management teams around a problem if necessary, on a weekend, at night—it doesn't matter. In a sense, when PAETEC goes live with telecommunications service to a customer it hasn't quite begun delivering the product. The real PAETEC delivery arrives when a customer calls us with an unexpected need and PAETEC service comes into play.

In this economy, the *ostensible* product—voice and data delivery—has become a commodity, like most products and services now. A customer can choose from a dozen nearly equivalent suppliers. As a result, what PAETEC offers, above and beyond all the competitors, is premium service based on a long-term customer relationship, any hour of any day. The NOC is where PAETEC answers calls from a distressed customer with a human voice—not a computerized call management system—usually within two rings. That first sign of trouble or confusion from a customer is when the real sale begins. We want outrageous levels of quality at reasonable prices. Most MBAs will tell you it can't be done. They'll say you can't do both. We do.

On average, PAETEC gets over a thousand calls a day from customers, for emergencies, questions about service, or simply the need for contact of some kind. That may seem like a lot of calls for a company dedicated to high-quality service, but part of the reason why customers stay loyal to us lies in the amount of care we put into our response to these calls. In the NOC, where the company solves these problems, service technicians have helped give PAETEC a monthly customer retention rate of 99.5% or better since it was founded in 1998. That quality of service grows directly out of an "Employees First" culture and the leadership it cultivates.

Our Number One priority is doing everything possible to make a customer's emergency *our* emergency too. Most of the time, and today with Trish Dow is a good example, we don't have to scramble this way. The system works smoothly, even with a number of outages.

Trish goes from one meeting to another, always checking her monitor and her voicemail to see if problems have escalated to the point where she will have to talk with customers herself, and as she works on the future of customer service—better ways to handle it—there's a steady stream of service conversations going on, behind the scenes, with customers and her reps.

Amber is finishing a service call:

"Let me give you my cell phone. Let me actually turn it on."

Samantha, the customer, laughs: "Oh my goodness, you're so funny."

"You have a piece of paper? I have the trouble ticket number."

"Yeah, a little sticky here. Let's go!" Samantha says.

Amber gives the number for the ticket. Samantha repeats it.

"You got it. We'll call you back in half an hour. A little snip snip and everything will be okay."

"Can I call you back in fifteen minutes?"

"That's fine. Call back whenever you want to."

The mood is tense in the Map Room, where our long-term thinkers are trying to anticipate the future of customer service, as reps handle twice the call volume of a typical day—friendly, unruffled, with no more than a few minutes of wait time for customers on hold. Today our long-term thinkers are worrying about the near future. They've assembled in a conference room with a globe and various maps on the walls, a fitting place to be taking the long view. But the long view, in this case, means a few weeks into the future, when we're planning to go live with VoIP telephones for 200 students at St. John Fisher College, in Rochester, and various people, including Trish, are meeting to talk about how we're going to handle trouble calls for those students when no one in our NOC has been fully trained to take calls on this new technology.

VoIP has arrived, but it's still new, as are most of our customers. It's already becoming familiar to people through companies like Vonage. Time Warner offers telephone service over the Internet as an add-on to Roadrunner broadband service. PAETEC is entering this market as rapidly as it can, but we can't, at the moment, hire qualified people quickly enough to handle the number of customer issues we expect, and we're going to burden already overworked people with something unfamiliar to them.

Jim Manetta doesn't look happy. He sits at the end of the table, occasionally answering his cell phone to stay on top of the outages —already a strain—while trying to face the fact that our introduction of VoIP at a local college could turn out to be the biggest headache of his life.

"I've got an outage going on so I'll be in and out," Manetta says.

Everyone agrees that Eric Campbell will be the only employee trained to understand and handle calls from students, but that two people already working in the NOC will work with Eric, learning as they go. Eventually, they will be able to handle the calls on their own. Eric, as if on cue, walks in: goatee, shaved head, long-sleeved V-neck sweater. There aren't any seats left, so he stands in the doorway.

"I'm 90 percent where I'll need to be, man," he says. "At the same time we're doing this with you guys, I'm doing a similar type of deal with engineers. The way it's shaking down, while I train your guys on how to troubleshoot, I need to train engineers on how to build it. It's not that easy."

"Holy moley," someone says. "Wow."

"Thank you," Eric says, for the recognition of his workload. "So, I'll be on call. Once we're up, if the customer is in trouble, that's my number one priority. When the trouble is solved I go back to what I'm doing."

"I'll dedicate as many resources as I can at all times. Today it could be an hour, tomorrow zero," Jim Manetta says.

Trish listens to all of it, taking notes, her foot waggling with nervous energy.

"We didn't make it to the moon until Kennedy said we would. Jim will be very challenged to provide resources," says Greg Utberg, director, switch operations.

"My repair team is the feeding ground for the NOC," Trish says. "We can identify some people to move in and handle NOC work, and we can take some of that burden off, if necessary." Everyone is relieved to hear this. It will help.

Then it's on to the next meeting.

"It's a great day at PAETEC, this is John, how can I help you?"

"John, we can't dial in or out and it looks like all our lines are down."

John asks: "What do you get with outbound calls, a busy?"

"I'm getting this. You ready?" The customer holds up the phone so John can listen to the recording: *I'm sorry all lines are currently busy.* "That's what I've been getting on outbound. If you call in you get a busy signal."

"I'm opening a high priority ticket."

He gets the customer's cell phone number.

"Call me back as soon as you know something," the customer says.

"It'll be thirty minutes."

By the afternoon, service has begun to come back up. Even with the call volume at double its usual daily rate, no calls have escalated to Trish's phone. The system has worked. Calls come in, they are routed to the NOC, work begins on the problem, and customers are contacted every half hour, or less, with status reports on their problem.

Trish Dow has been at PAETEC for seven years. She was our first customer service rep and order processor. She joined us two years out of college. She and her husband, Steve, took a year after they got married and camped in every national park in the country. They ended up in Portland, Oregon, where Trish worked at U.S. Bank, helping upgrade software systems. She and Steve returned to New York State, and Trish hoped to get a doctorate in biology, but they spent one day in Albany, where they'd planned to settle, and decided instead to come to Rochester. She saw an ad in the newspaper for job openings at a new company called PAETEC. She came in for an interview, and has been coming back every day since then.

There's almost no line dividing Trish's personal life from her work life. She is PAETEC.

I was out on maternity leave, but we didn't notify customers about that. About an hour or two after I had my daughter, Celia, I got a phone call with a Connecticut area code. I was in bed, and I looked at my cell phone and I figured it was my mother calling from her cell phone or something. I said, "This is Trish." It was a man who had a problem with the way his Internet service was billing on his invoice. I had just been to work the day before. I said, "OK, sure, I'll follow up." I called customer service and they took a look and fixed it. My husband was like: "Who was that? What did you just do?" I didn't even think about it. I never told the customer I'd just had a baby two hours earlier. I didn't want to make him feel bad. If I hadn't picked up the phone he would have left me a voice mail and I'd have called him back anyway.

By 5:30, Trish is cleaning up her desk and heading home. It's been the busiest day of the year for service—more than 2,000 trouble calls have come in today, with one customer calling in twelve times—and yet not a single call has been escalated to Trish, let alone to anyone above her in the organization. All in all, though PAETEC had never in its history had eight distinct outages in one day, it was, as one person put it, "The best outage we've ever had." She drives home and she keeps her cell phone in her pocket even while she gives her kids a bath at 9 p.m. But there's no need. She doesn't expect to handle any calls herself. Her people are on top of it. She spent two hours, in the middle of the afternoon, listening to the calls. She was amazed that everyone on her team sounded happy, energized. Not a hint of weariness or impatience in any of their voices. Her people take pride in that tone: it makes all the difference. One of her reps actually has a digital voice recorder, and she records some of her own calls and plays them back later to make sure her voice is as friendly as it should be.

WHEN THE GOING GETS TOUGH, CALL PAETEC

In the telecom industry, when you do something well, no one knows you're there. Telephone and Internet service are like tap water and oxygen: you hardly realize they're there until they go away. With telecom, customer service counts most when the system goes down. You don't see the true nature of your service provider until it stops providing the service: the most crucial test of quality is how a telecom responds when you aren't getting a dial tone, or your bill is wrong, or every incoming call gets a busy signal. And no matter who you sign with, something, at some point, will go wrong.

It may seem odd to devote an entire chapter to highlight all the various ways PAETEC customers have been frustrated by their service, but it's where the heart and soul of the PAETEC brand emerges. How we respond to problems—sometimes massive problems that, for a while, seem insoluble—is what keeps our customers with us. We have a list of customers who were without a dial tone for nearly two days during one outage, and they are still loyal to us.

Here, from a NOC manager, is one example, among hundreds, of what I'm talking about:

> Five years ago, we had one very very angry customer. Every night Verizon kept disconnecting the service, because of a fault in one of their systems. I worked for eighteen hours straight with Verizon and the customer trying to figure it out. The problem went away, and we still hadn't figured it out. We had to wait for it to happen again. But I built trust with this guy. He was from Boston,

and he was a crabby, crabby guy. You couldn't blame him. But he knew I had the determination to work through this thing to the end. It happened again that night. It was literally happening every night. I stayed until midnight and I worked straight through. He would have gone over the deep end if he'd come in the next day and had no service. It was a software problem that kept interrupting the digital signal, and we finally isolated it and fixed it.

It's the manner you use with customers that counts. You take whatever they give you, and you don't give it back. There's no group of people in this company more dedicated than the NOC, field service, switch techs and field engineers. Everybody cares. The attitude is, if I don't do it, it just won't happen, because it's so busy. Those guys are unbelievable. You'll see notes from people at 1 a.m. who are just tying things up so that the customer won't be disappointed. It's an amazing group of people.

Sometimes now on a holiday that angry guy from Boston calls just to say Merry Christmas. He calls whenever he has a question or a problem. He was swearing me to death the first day. Ever since then I've been a friend, rather than an enemy.

This has been the norm at PAETEC, since we began. We have dozens of stories about how we've responded to unhappy customers, sometimes extremely irate customers—justifiably irate—and turned them into loyal adherents. We had one customer on Long Island, near JFK airport, whose service kept going down for weeks, and then months. We were leasing Verizon's network to deliver voice service to this business, and Verizon was doing work on its lines in the road not far from the customer's premises. It was a big, involved project, and the work kept damaging the lines, and this went on for two entire months, so that the customer lost service almost every day.

Karen Dupke, a former officer who has left PAETEC to be closer to her family in Detroit, recalls:

After we isolated everything else, I went to a vendor meeting and took Paul Bloom, our director of service engineering. I sat with the customer, and Paul sat with Verizon. It was a hold-your-hand meeting. The customer said, "I know it's not you guys, it's Verizon." He said he went out and talked to Verizon workers in the street, and they told him, point blank, his service would be down every day, but they were union workers, just doing their job.

We couldn't resolve the problem. So, without charging the customer, we put in a redundant T1 line for him, and skinnied it down as much as possible and made it work. Paul showed him, physically took him out and showed him, how to change the wires over and make the second T1 work. They were billing maybe a couple thousand a month. He was a small customer, an import-export company. He had direct trunk overflow, so he was forwarding all the company's calls to his cell phone. Finally, his service came back on the original line, so he ended up with a redundant line in case service ever failed again.

I invited him to our customer recognition event at the Rainbow Room. Normally only very, very large customers would be invited. NYU was there. Pfizer was there. Stony Brook. Hyatt. And this small import-export firm. He was delighted.

As Jeff Burke, executive vice president, has said, "I've had customers tell me they get better response from PAETEC on another carrier's problem than they get from the other carrier."

CHAPTER 8
Lead at All Levels

Fʀᴏᴍ ᴛʜᴇɪʀ ꜰɪʀsᴛ ᴅᴀʏ ᴏɴ ᴛʜᴇ ᴊᴏʙ, PAETEC people are expected to take a leadership role for themselves and others. There is almost no limit to how much departments and people are expected to assume responsibility for the customer, and, as a result, for PAETEC's success. Employees understand customers come first because the company can't make a profit without them. Their attitude, their motivation, their enthusiasm—the feel of doing business with PAETEC that they, as a group, make possible—is the unique quality we bring to our customers. It's why customers stick with us. And they quickly come to understand this and try to live up to it.

Everything we do at PAETEC is designed to encourage leadership. It's built into the way we do business, including the way we compensate our people. One of our newest hires, a young woman who hadn't had much experience in sales, began making calls in one of our toughest markets: New York City. Her team did what sales teams have been known to do, from time to time. They gave her an account they were certain would be impossible to crack: the city itself. City government. They told her to see if she could sell our PINNACLE software product to the city. She did. It's one of the largest sales we've ever made at PAETEC, and her commission was quite large as a result. Because she landed the account, she'll be collecting a small percentage of anything we make from New York City for the life of our relationship with that city government. That's the way it works in sales at PAETEC. That may seem unfair, someone starting out with a residual payment that lasts as long as she's here. But here's the question: how much leadership do you think she'll bring to the job of keeping New York City happy? How much time and effort do you think she'll apply to making sure our service to New York City is exactly what that customer wants and needs?

If you want to know the answer, look at PAETEC's numbers on customer loyalty: we have a churn rate of 0.4 percent. That's the lowest in the industry. We just don't lose many customers once we get them.

Inspiring everyone at all levels, to behave as leaders is the most difficult challenge any manager will face. Most people don't have the self-confidence to think of themselves as leaders. Yet that confidence can be instilled by being absolutely clear with everyone about a company's values and its mission. When those become internalized, small daily acts of leadership begin to feel right: they either fit into the mission and values or they don't. Our chief operating officer, E.J. Butler expresses it well:

> If you sit down with people and explain why you do the things you do and invite them into the conversation and make it collaborative ... people will understand what to do. At the end of the day it's not about what you pay people, it's about how you make them feel about working for you. People come in and ask me how they should feel or how they should act. They're asking me to tell them what the company values are. "You want me to be a leader? Tell me how." Instead, you drill them on the values and the mission, and, eventually it isn't what I say or Arunas says that counts, it will be what the branch managers say.

> When my four-year-old son goes into the backyard I have to be with him. At some point I'll have to get used to looking out the back window every four minutes. Then it will be letting him go completely. How do you do it? You go out and work with people and it's hard work to make sure they're emissaries of our mission.

One of the boldest ways in which PAETEC has created a process that instills leadership—and encourages empowerment—is the way it pays its sales reps. The company gives reps a share of ongoing revenue from any customer a rep brings to PAETEC, for the life of the business relationship. This inspires reps to stay in constant contact with customers, testing for quality of service, and offering ways to buy into a higher level of service. It's a costly policy that might strike some observers as unfair to employees who don't enjoy residuals, but it properly rewards those who generate crucial results and shows how much the company respects those results. It's a risky, high-turnover part of the company, and those who endure in our sales organization also endure some of the toughest battles. Some top sales reps make $8,000 a month before they sell anything, as long as the customers they've brought on board are still with the company. There's a catch, though: the sales reps have to keep making their targets to keep getting their residuals.

Most businesses don't compensate reps for ongoing business this way because they think it's too costly; the reps make too much money and will get complacent. Maybe so in other organizations. Not here. We've found the opposite occurs. Do they make very large incomes? Clearly they do. But we find them becoming incredibly customer-oriented.

Being focused on existing customers creates empowerment and leadership at all levels in the sales organization. As Dick Ottalagana says: "Once you give those people the ability to make the decisions and be accountable and raise their hands and say, "It's me, I'm responsible for our success," their thought process is different. "Now it's 'mine.' Not just PAETEC's." Self-esteem rises. That outcome, ultimately, is what we call "sticky business." Our customers don't want to let go of that kind of commitment—because whenever they interact with our people they can tell how much our people care about keeping their business.

Dick goes on:

> Regarding our back office systems: we make investments for the long term. But we were democratic in the way we chose them. We went around the table saying, "Tell me what to buy. We'll get the money." We had buy-in. We overspent in the first two years, but we had capacity in place. We believed we were going to be successful. So, when we put in the backbone of our computer systems, we overspent, and you wouldn't need to spend that kind of money with the customers we had, but if we had huge growth—and we did—to go back and retrofit would have been expensive. We did expensive scalable systems from Day One.

There is significant risk built into this way of working. With that kind of money at stake, the typical reaction of top management would be to seize control of the process and dictate what the company would buy. Not at PAETEC. The company runs on three central software programs: one for finance, one for billing, and the central program that forms the backbone of the company's internal operations.

We chose a Canadian company, Eftia, to create this central operating system. Our people considered another vendor, Metasolv, which was conventional, stodgy, not as flexible and not as cheap. But after a year of development, about the time we were ready to ramp up and put in our switches and go to market, Eftia was still struggling to deliver the software. So we quickly decided to switch to Metasolv and patch together what we needed until it could install the software.

The term "patching together" doesn't quite capture the state of emergency we were in. Eftia had failed. They just could not deliver. At the start, their product demo was wonderful. But when it came time to deliver, a few days before we were to get the system up and running, their senior team was off to Mt. Kilimanjaro for a retreat. We fired them and went to Metasolv.

There was no wiggle room for anyone. The company's fate rested on a successful launch three months later.

We needed to implement in 90 days. We took nine director-level people right out of their jobs and dedicated them full time for three months to get the system up and running. And it worked. Dick remembers: "They moved a mountain. Metasolv could not believe we did it. Usually something like this runs downhill. But after all, we'd empowered our people to make the choice, so we could empower them to solve the problem. We had only a handful of customers at this point, but we were on the cusp of huge growth. We had some billing but even that was messed up."

There are no fire hydrant painters who refuse to fight fires at PAETEC. When the building is on fire, everybody grabs a hose, even the guy who's job is to hold a paintbrush. No one says, "That isn't my job. My job is painting."

The new system ended up working perfectly, and the company stayed true to its employees-first philosophy when it counted: it had empowered them to choose our software company, and in this one instance, they chose badly. The company lost money, and its people lost sleep—and some hair. But it didn't lose any business. Management didn't line up scapegoats and fire people. This may not have been consistent with the accountability inherent in empowerment, but management accepted the inevitability of mistakes and refused to monkey-manage the problem down onto the shoulders of people who wouldn't be able to handle it. It mobilized, put the hydrant painting on hold, did what needed to be done for the sake of everyone, learned from it, and moved on.

Metasolv said it was the smoothest implementation ever, in their history as a company.

INNOVATION AT ALL LEVELS

From this level of service and support comes a new kind of innovation. Our people constantly come up with ideas on how to improve what we do and what we offer. And from this attitude—whether or not we use the ideas our people suggest—arises a willingness to own up to mistakes, flaws, and oversights, because once you find a flaw you can distinguish yourself by coming up with a way to make it better. Mistakes become a prime opportunity to shine.

From its first day of doing business, PAETEC's work processes were designed to qualify the company for ISO (International Standards Organization) certification. Recently, we've also added Six Sigma processes. Embedded in these quality programs is the implicit assumption that all people in an organization will continually improve the way they do things. For example, Six Sigma provides tools for analyzing any process in a company, measuring it and devising ways to improve the results the process generates. Anyone in the company can apply these tools from day to day—formally, in a group, or informally, simply as a mindset aimed at improvement. As part of this effort, PAETEC has empowered its employees to be creative in the way they satisfy customers—to take responsibility for offering only the highest quality.

We expect everyone to focus on continuously improving every process at PAETEC. That means being ultra-attentive to when our service falls short. On the service map stretching out over the screen of the huge, two-story monitor in our Network Operations Center there is often a switch component that isn't working, something going wrong—the display makes it inescapable. If we're alerted to a problem, we immediately spring into action, and often we can correct the problem before it even has a noticeable effect on a customer's service. Something will always be breaking, going down. Being aware of it, acting on it, having a plan for how to manage it and then working with the customer constantly to restore proper service—this is how PAETEC has its best chance to shine, to show how much more it can focus on a customer's needs, compared to any competitor.

Departments collaborate freely with one another so that organizational learning and innovation rise up naturally from lower levels if the organization, rather than being imposed from the top. Jason Elston, who manages the NOC, says, "One thing that endeared this place to me from the start was that you could have a good idea, and that good idea could become company policy." The culture of cooperation, of knowledge sharing—all the ways respect for employees creates work processes that give people the freedom to learn and decide and take on leadership—helps generate products and services as a function of doing daily business, not as a separate research and development operation.

It's assumed that workers know what needs to be done better than anyone who manages them. At PAETEC, the motto is, in President Reagan's old formulation about the Soviet Union: "Trust but verify." You are presumed a hero until proven otherwise. Mistakes aren't just tolerated—they are seen as a prime opportunity for growth and improvement. And innovation that leads to improvement is rewarded.

The kind of commitment to improvement I'm talking about isn't something you can instill through processes and recognition programs. You *hire* it and then you nurture it. Again, the central message of this book: what matters is the character of the people you bring on board. Either they have that PAETEC quality or they don't.

Scott Rubinson, director of account development in New Jersey, put it perfectly:

> We choose to come in and do this every day. You have to want to come in and never slack. Account managers never want to just get by with what they can do. You have to go that extra mile for the customer. Do you get comped on what you do, afterward? It's secondary to doing the right thing up front. We are good as a team environment. When there have been service outages, I have gotten calls from other account managers: "You are getting bombarded, can I help?" This is unheard of in most places. They get no reward for helping us out from where they are.

> PAETEC gives you enough information to do your job but it tasks you to learn and grow on the job. We don't go out and fill positions. Often, we hire go-getters, who have a strong work ethic, and we figure out a job for them. We ask a lot of every department. In one merger, we inherited seven account managers, and only two are left. It wasn't because they were bad people or didn't know the job, but you can't drive twenty miles an hour where the average PAETEC speed would be sixty and most are going eighty. You aren't going to make it. We work at a fast pace, in an environment where you multi-task, and do it all pretty well, go home and do it again, the next day.

> We expect people to think on behalf of the whole company. They give us one statistic that encourages that kind of thinking: revenue per employee. Is it up? Is it down? A fascinating stat. And it makes you feel as if your effort is having an impact, if the number is up. Here's the number of employees and here's the revenue. It's the coolest stat because it shows you how to make money through productivity. Everybody sees it internally. I know what it is right now, actually. It's still $400,000 per employee per year.

As Scott points out, leadership—and the superlative service it creates—grows out of the interaction between the right group of people allowed to work in an organization that gives them the knowledge and the freedom to assume leadership responsibilities. One of our trainers says, echoing others in this book, "You can see the impact of what you do. That's one thing I like about this company. It's small enough that one idea can still affect the organization. I could go up and ask Arunas for five minutes and lay some whacked-out idea on him, and he might go, 'Are you crazy?' but he'd listen to me. At least you get your audience."

WE STAY IN FRONT OF THE CUSTOMER

We don't keep company with our customers only when they have a problem, though. Great support and service means staying in front of customers as often as they'll let us. There's no substitute for face time and familiarity. You can't build a friendship without it. The way we like to put it is, in football, nine times out of ten, the team that has possession of the ball for the most minutes will win the game. If you have the ball for forty minutes out of sixty, you're just way more likely to score. We're small enough to have high-level people in front of the customer far more often than, say, Verizon or AT&T. With those companies, the customer just won't see an executive. It works all down the line. We're constantly staying in touch, checking to see if they're happy, looking for opportunities to upsell. Our reps are just always there. And when they call, they hear a human voice, not a list of choices for which button to push. In other words, we try to possess the ball more than any other competitor.

Not long ago, I drove to Hartford to meet with customers, when my flight was grounded because of high winds. I met with people at Pratt & Whitney, Chubb, and the University of Connecticut. The meetings were set up, and I could have troubled these people to reschedule a new time for me, but I believed it was better to show them how much they mattered by doing the six-hour drive from Western New York to Connecticut.

No one would have thought bad of us if we hadn't gone, but they were very impressed at how much effort we'd put into showing up, and—this is the real point—it was a chance to see them, face to face, which we never take for granted. We only have eight people in Hartford. They work hard for us. It's an outpost, like Fort Laramie, back in 1870s, with a couple settlers. So, to support our own people, and to show our customers how much we care, Lisa Fitzgerald, Jeff Burke and I went down in one car and visited everyone. We even got a request for us to answer a request for proposal—because it was clear they wanted to do business with us by the end of the visit.

One day, I was going down to say hello to the manager of customer service, Lisa Chapin. I walked in on her and her supervisor in a conference call with customers. I listened for a while, and understood what was going on. Service to one of our customers had gone completely down, and he was on his cell phone. Three seconds after the call rang on our end, we picked up, and within a few minutes, we discovered the problem was with our carrier, another telecom. Our customer had gone through twenty minutes of automated systems with this carrier and was getting bounced around. We asked, "Have you talked to somebody in those twenty minutes?" He said, "No, I got disconnected three times." He finally got to a switching location which was the source of the problem. All of this telecom's services were down. They were trying to troubleshoot the problem but couldn't figure it out. Meanwhile, when they put him on hold, he was talking to Lisa and she was saying, "From what you've told me, I know what happened. They actually shut you down."

He was scheduled to move locations, and he'd called to delay the move, but someone at the other telecom had mistakenly shut him down, thinking his move was in progress. The message hadn't gotten through to this person that they were supposed to delay the shutdown. The guy said to Lisa, once she had explained this to him, "I'm giving you my local business. You actually picked up the phone." Lisa said, "You know what, I'm going to have one of my account management people give you a ring and tell you what our services are, and you just stay with me, and I'll follow you through the whole process."

Almost everything we do can be reduced to something very simple: when something goes wrong, you'll hear a friendly, helpful human voice on the other end of the line, after a couple rings. It's all built up on that foundation.

One of our VPs in New York City reflects on how much our people, and our customers, appreciate this sort of support from executives in the company. During a sales blitz, or during an informal visit like the one to Hartford, PAETEC sales people almost expect that one of us on the senior team will show up to talk with customers

> Arunas just jumps on a plane and will fly in to take customers out to dinner with my reps. The senior team wants to get to know everyone. They meet with all the reps, they sit down, talk, listen, care, and ask advice on how to do the job better, and then they go out and meet with customers. Customers are impressed that they're talking with the CEO and his top people.

Wendy Showers has on her office shelf a crock bearing the words "Ashes of Problem Customers." It's a joke, but not at the expense of our customers. The crock, of course, is empty. There's really no such thing as a problem customer at PAETEC. We have said goodbye to a few, without regrets, but we consider it a privilege to serve even the smallest and most difficult customers. Wendy points out that staying in front of the customer, keeping in touch, doesn't have to be done in person:

> Many customers don't care if they get an account manager meeting with them face to face. The biggest thing we've found is that they want their account manager to pick up on the first ring when they call and need something. Out of our LA advisory board, there was one particular customer who said he didn't want our employee wasting two hours on the LA freeway to ask him how he was doing. He'd rather that person was at his or her desk to pick up the phone when he calls. Nine out of ten times it's just a question that needs answering. Someone who can make two face-to-face appointments per day could do twenty of them by phone.

This is simply something we've learned how to do. It's a fine art.

WE LET THE CUSTOMER BE THE LEADER

Great service and support also means making self-service easier. PAETEC customers can actually manage their own accounts by going online and delving into the details of what we do for them. PAETEC Online offers every customer an individual portal into a complex, sophisticated record of service. By logging onto PAETEC Online, customers can unlock a powerful, secure source of information to track PAETEC services, what has been spent, and how we might modify our services.

It's essentially an online customer service representative. It enables a customer to analyze service usage patterns, look for trends before they become problems, and plan for ways to alter the level of service we provide in order to adapt to changes in needs. It lets a customer manage PAETEC services by adding or changing features, reviewing bills and payment history, or resolving problems. Most other integrated communications providers in PAETEC's class don't offer this kind of communication with customers: it's personalized, it gives immediate insight into the inner workings of what we do for each customer, and it offers a greater degree of control in the customer's relationship with us.

Eventually we want to personalize PAETEC Online, so that when a customer logs on, the page will reflect that customer's relationship with us, the way Amazon.com offers suggested books based on purchasing history.

Our goal with PAETEC Online is to get it to a point where PAETEC is like the bank and PAETEC Online the ATM machine, so that customers who learn to use it would rather go to PAETEC Online than call customer service to handle routine activities.

To fully utilize this kind of service, a customer has to have a clear and strong sense of what PAETEC offers, what to expect from the quality of our service, and how much control he or she can have over that service. So in 2006, we offered a new customer onboarding process to help orient new customers to the PAETEC experience. As soon as you've joined PAETEC, before service is cut in, you will receive a welcome kit via email. It contains everything a customer needs to know to get the most out of PAETEC service: basic information about the company, a map of where we provide service, our basic product and service offering, a tutorial on how to get the most out of PAETEC Online, and an escalation list for customer service. The sales office assigned to the new customer will customize each kit with additional information about service in that customer's region. Not long after this initial mailing, an account rep dedicated to that customer will pay a visit to deliver his or her personal cell phone number, which the customer can call at any time of the day or night, to resolve any kind of issue with the company.

As part of the onboarding process, the customer will receive a satisfaction survey six months from when they joined us, and twelve months after that. The process helps both the customer and PAETEC to create an open line of communication, so that customer and company share knowledge essential to providing the best possible service: it helps all of us get to know one another.

Based on this knowledge, our database will customize the messages a customer receives. As part of the process, a customer will indicate how he or she wants to receive information from PAETEC: by email, direct mailings, or phone. The system will have a set of "triggers" for action: if we hear from a customer about some area of interest, or get a request for more information about a product or service, we'll respond quickly to satisfy the need. The process helps us keep a clean, accurate database, with the most useful and up-to-date information about our customers, enabling us to direct only the most welcome kind of marketing to individual customers, based on the preferences they've given us.

OUR DEDICATION TO CUSTOMER SATISFACTION IS PERSONAL

We devote three key people to the happiness of our customers: the direct sales rep who brought them onboard, the account manager assigned to them, and the account engineer. Once you join PAETEC, you can expect your rep to call on you to make sure your initial billing is accurate, service is superlative, and cutover issues are resolved in a personally satisfying way. The rep will also explore whether or not other services and products offered by PAETEC might better fulfill your needs. The idea is to have one key person dedicated to a customer's satisfaction, available at any time to respond to any issues that might arise. It doesn't take long for that relationship between rep and customer to make doing business with PAETEC a more personal experience.

Supporting each sales rep is an account development manager, whom the customer can call at any time. He or she serves, internally at PAETEC, as the customer's advocate. Account managers call on customers regularly to stay in touch and answer questions about order status, when new services will be up and running, measures of performance or almost anything else related to the account. The account manager is the prime contact during network troubles, the liaison with our Network Operations Center. Working with the customer's dedicated account rep, the account manager helps craft personalized solutions for each account's budget and business strategy.

Assisting the sales rep and account manager is the engineer assigned to an account, who becomes intimately familiar with the technical demands of the customer's system and service. A key part of the engineer's job is to explain to customers the importance of diversity in service, because customers should have backup systems in place to maintain service when problems arise.

We also offer free diversity audits for all customers. These audits are fact-gathering sessions to discover whether or not, during any problem with service, the customer has an alternative way to handle the traffic essential to business. It's our way of encouraging customers to create redundancy or fallback systems for handling telephone and Internet service when problems with the primary system arise, or service is interrupted by any weather or power failure. When we've completed the audit, we suggest a sufficient disaster recovery plan based on the particular needs of each individual customer.

CHAPTER 9

Respond with Care and Character

W HEN YOU'RE TRYING TO ACHIEVE unmatched service and support, you need to keep things personal. Our customers are our friends, and we depend on those friendships to stay in business. You don't say to a friend, "I know you're unhappy, but I don't feel obligated to come over and talk about it." You *want* to go over and talk about it, and then do something to help.

An East Coast sales rep expresses it this way: "It would seem logical to think, 'This isn't personal, it's business.' But you have to keep it personal. If a customer needs somebody to yell at, that's who you are. You can't take the customer's side to the point where you badmouth your own company, though. You're always the customer's advocate at PAETEC, and a lot of your task is to pass on the customer's passion and frustration to other PAETEC employees. Yet you have to do the same thing the other way, too. It's a delicate balance where you have to deliver a message that supports your belief in your own company to the customer, without telling the customer it isn't our fault, even if that's true."

He goes on:

I remember the first big PAETEC outage. I was on a sales call, and the company's office manager came flying into the conference room. The door just flew open, and he said, "We have no phones." At the same time, my own cell phone and pager were going nuts. I knew something very bad was about to happen.

The Philadelphia switch was out in the only part of the switch where we didn't have backup. Once we realized there was an outage, I made a couple calls from there and we suspected it was widespread. That particular customer was so new that I was there to review the first invoice. The service had been up for less than a month.

We found the outage and discovered how many lines were down. The timing card went out, and we didn't have another card there at the switch. For a couple hours many customers served by that switch were down. It was after hours, so most businesses didn't realize it. The 24-hour shops understood and had enough redundancy in their plans, on their sites. In this case, it was our fault—we didn't have the part we needed there at the switch. We told everyone that's what had happened. When you're honest with customers and keep them abreast of your progress, it makes a huge difference. We made sure their backup line worked. They were running a law firm with one line. It was the end of the day, so we lucked out.

Every single customer in my base was down on that one. We had VPs in the NOC to talk to customers who needed it. We cared, so we got through it. We ended up losing no customers.

IF IT'S YOUR PROBLEM, IT'S OUR PROBLEM

A power outage isn't the first hazard you think of when you imagine one of the year's four major golf tournaments. But when the PGA Tournament came to Oak Hill Country Club in Rochester a few years ago, much of the Northeast region of the country lost power, and though it didn't affect play, golfers were lighting candles in the locker room.

It was after the main shift, but PAETEC scrambled. Within minutes, the NOC was full of data engineers and system engineers. Our NOC has emergency backup power so we were well suited to help customers, on their cell phones, transfer their traffic to other lines or offices. We didn't need to persuade a single employee to come in. When their power went out at home, they were like volunteer firefighters. As soon as they recognized the crisis, they jumped in their cars, and they were here.

A manager in the NOC recalls:

Not only my own techs were here, I had other groups asking how they could help. We assembled on Thursday night, with a full staff until midnight, and all management was ready to be paged or called. We moved into Friday the same way, working from about eight a.m. to midnight Friday. By the weekend, we were comfortable, even though so many people were still without power. It wasn't sensible to start working on their phone issues at that point: all the carriers were so backed up. But we could status them. We didn't do anything until power was back up at the site of the problem itself. As a management team, we scheduled three calls a day to say this is where we're at. Everybody rallied, everyone was on the same page.

Another NOC manager from that crisis recalls: "Did I tell those people to stay? No, I told them to go home, because I needed them in the office the next day. They said, 'Don't worry about it. We'll go home when we're ready. Just leave us alone.'"

The heart of great customer service isn't simply the ability to fix a problem quickly, although that certainly helps. It would help keep customers happy, of course, to solve every problem in fifteen minutes. But telecom, let alone the rest of the world, doesn't work that way. But by the time you are finished reading this chapter, you'll realize that what matters most is *how* a company treats a customer with problems, not always how long it takes to solve the problem. What matters is a gradual accumulation of dozens of small things: the way you respond, which means the tone in your voice when a customer yells at you, how quickly you pick up the phone, whether you remember the customer's name, how often you call the customer back to let him or her know what's happening, and how well you listen. These aren't courtesies unique to telecom, or even to the world of business: they have to do with respect and caring. Human beings appreciate being treated with civility and patience and understanding.

One of our executives puts it this way:

One thing I say to vendors and customers is that we answer the phone. We make sure your call is answered and that your problem is understood, and we make sure to call you back. I want you to be informed and told the truth and understand the effort we're taking to get the problem resolved. You can't ask for anything more. We give our customers frequent status reports. Within thirty minutes after the call, our goal is to give a status report. And every thirty minutes after that, until the problem is solved.

Almost two thirds of our trouble tickets are not our fault, but it's our customer's plight, so we take responsibility. We close our trouble tickets with a phone call live with the customer: "Are you happy, is everything OK?"

Another NOC manager says:

Almost all of the game is telling the customer what's up. It's almost better to call the customer and *not* work the trouble ticket than to work the ticket and not call the customer. This company has changed not only my outlook on a lot of things, but my family's too. We bought a dryer from Sears, and Sears called us before we could call them, and asked us, "Are you happy with your product?" "Yeah, that's Sears," my wife said. My thought was, "That's PAETEC." I don't take that for granted. That's what makes Sears different. And it's what makes us different.

Many times, we show up at our customer's door, even if there's nothing we can immediately do about the emergency. This was true in Manhattan during 9/11. One of our people, Rosemarie Tolins, showed up at a customer site on Wall Street and sat with them, helping out, for three days. She couldn't *do* anything directly, at the time, to help restore their service, but she wanted to be there, offer her emotional support through the crisis, in any way she could. Again, this was not a big customer: they were billed around $1,500 a month. She was pitching in, answering their phones, doing whatever needed to be done. She was there for them.

During one crisis, also not of our own making, two of our top people were in a conference call for five hours, on hold waiting, until they were able to get the vendor, whose equipment was causing the outage, at home. When they finally reached him, one of them had been sitting in his driveway for more than an hour, with his cell phone plugged into his cigarette lighter, waiting for the call to go through. He didn't want to get out of his car, because he was afraid, after all that time, unplugging his phone might somehow disconnect him.

One of our NOC managers recalls a situation where one customer was having intermittent failures on a T1 line, and the equipment wouldn't restore itself. Something was taking it down in such a way that technicians had to manually intervene and reboot it. In order to make sure it wasn't our T1 line causing the outage, we put in an entirely new line, without charging the customer. Then the same thing happened on the new line.

He says:

Most companies would say, "I've put in a new line, and I see the problem is on your end. There's nothing I can do." We didn't. We said, "OK, let's get one of our most intelligent PBX guys and see what's going on." He came to us when we acquired CampusLink, and he knew PBXs. Jim Dancer. We asked him to focus on the problem. So he took it as a project and worked on it. He made the effort to get with the manufacturers, and he worked with technicians on trapping calls and trapping failures. He gathered information on the status of the T1 when it was up and when it was down. It took him a couple months to capture all the information on the event and work this out. Ultimately he determined that it *was* the PBX. The manufacturer finally said it was a known problem, and they installed a new patch, that would enable it to reboot on its own. The customer put the patch in, and it's been fine.

He recalls another instance, where we kept a customer happy by doing something above and beyond.

We were having intermittent drop-offs on a customer's T1. We did all the testing, and we found it was the Network Interface Unit, which connects our T1 to the customer's internal network. There was one specific NIU card that kept the circuit stable, but the billing system didn't work right. We put in a new NIU card and everything worked. It wasn't our card. The customer wanted all his NIUs changed at all his different locations. That was not an easy task. Again, none of this was our problem, nor our responsibility, but we went through and changed all the NIUs. And we charged him nothing for it. The point is the long-term relationship, and the customer's satisfaction. It doesn't matter who's at fault or who's responsible. The one who steps up and takes responsibility is the one who'll still have the business five years down the road.

The only disadvantage to this is that our customers learn that this is the way we work—that we take responsibility for their happiness, even if it involves someone else's equipment, or network, or service. So that gives them the ability to take advantage of us, when they don't have any other choice. We've had NOC technicians get a call to say a customer's data service is down, and even though it isn't our router giving him a headache, because it's the customer's core business, we will go in and fix it. In one case, we sent two people out, and they responded to a problem like this, and they fixed it. The only reason the customer called PAETEC was that the person didn't know what else to do and couldn't get the router's actual vendor on the phone. It took our people more than half a day—getting there, doing the repair, and getting back—but that's how we respond to a customer who puts that much trust in us and believes we can help. It will come back as a reward some day because it encourages customers to stick with us through bad times.

Another vice president remembers another instance just like this:

I had a senior account manager in New York who discovered one of our engineers at a customer's site didn't have a piece of hardware to cut in service for a new customer. The senior account manager went to the office, and got the hardware and delivered it. It was a two-hour drive, to Westchester County, where customers often aren't that close. Two hours out and two hours back for that. This is a *senior* account manager.

THE BOSTON OUTAGE

The biggest network failure PAETEC ever experienced began on Saturday, Dec. 18, 2004 at 7 a.m., in its Boston switch. It was a potential disaster. One tenth of the company's customers in that area were deprived of some telecommunications services, and many were completely down. It was a harrowing break in service for some customers who could least afford it: hospitals, ambulances, and other emergency services. At the first sign of trouble, the company mustered all of its regional management on a conference call.

At that time, one person needed to take the lead. Ultimately it fell to Raffi Yardemian, director of account development in Boston.

At the time, there was only one person in Rochester in the NOC by himself. So when the phone started ringing off the hook, all he could do was take down names and numbers. But he couldn't keep up with the flood of calls. The phone just kept ringing and ringing. At the time, we didn't have a provision for an answering service to kick in when this happened because we had been so focused on having a live voice answer every incoming call. In six years of doing business, the company had never had an outage of this magnitude with only one person to answer the calls for help.

Understandably, customers became angry and baffled, partly because their expectations were so high, based on past service. Nobody was picking up the phone? Unheard of. It was the opposite of everything PAETEC stood for. We boast to customers that we'll answer calls after no more than two rings with a helpful, human voice. In fact, someone was answering, but it was only one lone NOC technician, who couldn't even bring himself to say, "Hello, it's a great day at PAETEC," before he asked for a name, phone number, and then hung up, so he could answer the three dozen other calls coming in. Whenever he could, he'd call Raffi and say, "I've got hospital X, down they need a call back. I can't even open a trouble ticket. I don't have time."

We sprang into action. Raffi delegated a number of calls to managers or people in the area, mostly to himself or one of seven other managers. At end of the first two hours, he added it all up: eighty customers needed follow-up. So he assigned ten customers to each manager he'd been able to mobilize. They broke it into two groups: those who called customers and told them what PAETEC was doing, and those who kept tabs on what was going on technically to solve the problem.

This summoning of resources lasted two hours: each manager would pass a call along to a sales person who would call the customer back to talk about the problem and what was being done. Mostly, customers just wanted somebody to talk to. Raffi stayed on the conference call, asking his people to call back and let the NOC know what level of emergency each customer was facing.

By now, the technical team was at the switch. With growing despair, the team was beginning to realize it had no idea why the switch was down. The system wasn't able to route service to those who needed it most, as it was supposed to do. The system would have forwarded calls if it had recognized there was an outage, but for some reason, the switch wasn't sending a distress signal to our network system. As far as the system could tell, everything was fine, even though this was far from the case, since at least eighty customers were without all or part of their service.

Alert Ambulance in Boston was probably the most irate customer, but there wasn't much we could do for them. They were completely down for inbound calls. They all had cell phones, but we couldn't reroute calls to them because the system didn't recognize the outage. So we split up everyone into two conference calls, with our operations team headed by E.J. Butler. The rest of us, working off the first conference call, headed by Raffi, were talking with customers, giving them the latest news. E.J. would jump in and say, "Here's what we're doing, and when we should be back in service." Very quickly, it became obvious these words didn't count for much. No progress was being made. No one could figure out why the switch was misbehaving.

For the first few hours, our customers appreciated the phone calls, but as it became obvious no progress was being made, they got more heated. Five hours into the outage at noon the break in service seemed even more mysterious than before. Three technicians were working feverishly at the switch, and most other PAETEC people were working at home, either on conference calls or talking with customers. Back at headquarters, operations people were doing remote diagnostics. Everyone believed the outage had something to do with an upgrade to the switch that weekend, a new patch which had been installed. But Lucent, who'd made the switch, said they'd never seen a problem of this magnitude. Nothing had changed. The switch wasn't working, but the system wasn't recognizing the problem, so there was no report to show what customers were being affected by it. If customers called in, we knew they were in distress. Otherwise, we could only guess who had a dial tone and who didn't.

So it went into the afternoon, and nobody knew what was going on. The switch module, which was down, potentially affected hundreds of customers, as many as seven hundred T1 lines. At 3 p.m., we decided to revert back to the system we had before the upgrade. We thought it would all come back up. An hour and a half had passed when, Lucent said, "Let's try this maintenance at 2 a.m.—take the whole switch down at an hour when it will have the least impact."

This is how bad it was: we were anticipating additional hours of down time, and just planning ahead for it.

Around five o'clock, Raffi called all account managers and told them what was going on: every half hour or forty-five minutes somebody was calling the customers who were out of service. Alert Ambulance was furious with PAETEC at this point. The man responsible was heading out to a holiday party, where he would be facing his CEO. He said to us, "I may lose my job, and you can't even forward our incoming calls to my cell phone?" Raffi said, "Have your CEO call me and yell at me."

As Raffi recalls: "I don't know how that guy was able to talk to me, I would have been so upset in his position. I told him what would be going on at 2 a.m., and we didn't hear from him again until midnight. He called and asked if two o'clock was still on. I told him it was. He still had his job."

It's a hard task to tell people who have already been down for six hours that they'll have to wait longer. Raffi said, "I know you'll be upset. But here's the story. I've got thousands on the switch, and I can't take them all down now in order to help the others who have lost service."

Two a.m. arrived. Everyone was on one of the two conference calls: operations and customer support. Yardemian had a phone to each ear. Lucent was ready. The technicians at the switch had taken a hotel across the street to get some sleep, trying to get four or five hours before 2 a.m. rolled around, and now they were up, drinking coffee. But before they pulled the plug on the switch, one of them noticed something.

Raffi was listening on his end:

"I've got the Lucent guy here. I think he thinks he knows what the issue is," the technician said. "Why didn't diagnostic pick this up? The cables are plugged into the wrong holes. When they did the upgrade they plugged two cables into the wrong holes."

"But nobody was certain. Everyone talked it over, for more than two hours. At 4:30 a.m. they were still talking about slowly taking this out, slowly doing this or that. We went ahead and changed the cables, and it took a few hours to replug everything the way it should have been. By 6 a.m., we thought we were good but we wouldn't know until we talked to customers, because according to our system the trunk lines were never out of service—there was no indication of a problem on our end."

By noon on Sunday, PAETEC was certain it had restored full service, with a few intermittent issues here and there. But the customer care work had only begun. An intensive follow-up campaign began on Sunday. Raffi had a list of thirty customers with the worst lapses in service. Our COO took seven or eight of the calls. The rest were handed off to senior people. Raffi assigned them each a list of customers: "Here are your accounts and what their issues are."

Afterward, remembering this period, some customers said, "I can't believe you got your COO to call me, but thanks. I don't think I needed that, but I'm even more assured." Between noon and four o'clock, we decided which customers were good and which needed follow-up. We made it mandatory that all account managers come at 7:30 a.m. and start outbound calls to customers, after a brief update of the events of the two days. Everybody made it in on time. Two of our top people flew in to Boston to see four critical customers each, and it got interesting.

After all that down time, customers knew what we'd done and they actually sympathized. Someone at Roger Williams Medical Center, in Rhode Island, said, "I'm sorry, I guess it was rocky, but I'd have hated to be you guys, the amount of calls I got from you, man."

As Raffi puts it: "She was feeling sympathy for *us*."

Raffi sends out an updated escalation list with phone numbers every two weeks—it shows customers a list of people, of higher and higher rank in the company, they can call if they aren't getting the service they want. The latest list had just arrived at one customer's office, only days before the outage. In the cover letter, it said, "Dear PAETEC customer, we're having a maintenance session on such and such an evening." So we'd told every customer there might be a problem, in advance, and she remembered this. Despite all the down time she's been through, she thanked us for having actually notified her it might happen, with a list of people to call if she had a problem.

We sent people to see Landmark Medical in Rhode Island, a new customer. They were frustrated and upset that the NOC hadn't been able to take a trouble ticket efficiently. "You give us this list, but you can't take a ticket?" We didn't get the information to call them, so it took us three hours to get back to them. Raffi looked them in the eyes and said, "If we weren't dedicated, all four of us wouldn't have come out here. We want to hear your suggestions and tell you what we've got in the works to make it right." I had three of our people with me. The customer calmed down quickly and commended us on how we handled this, and is still a PAETEC customer."

We sent two sales managers to Alert Ambulance. We expected to tell them, "Yes we'll let you out of the contract. We understand." E.J. Butler said, "I can't talk with this guy anymore, he's very upset, he needs to talk to somebody else." So we sent another group. We knew we couldn't write them a credit as large as they deserved, so we were expecting to just part ways. Raffi told the people we were going to send out, "Go in there, fall on your swords, and say we don't credit business downturn or lost business, but we will help them switch to another carrier and give them a full month's credit." The customer said, "The way you handled this shows me you're the right carrier. I would like a credit and to stay with you. A $300 credit for vendor fee." As Raffi says, "It was amazing. We didn't expect to keep that customer."

Our findings were, because of our proactive approach, we had to give only $38,000 in credits. We figured, at the start of our post-emergency follow-up, if we came out below $150,000 we'd be lucky. We lost no customers. We may lose business eventually. One customer is giving us a reference about how well we handled the whole thing. The email to us said, "It's unbelievable what you guys have done. It's over and above what any other company does. It's the best company I do business with, not just telecom."

After it was over, senior people had spoken to a customer who had decided to leave us after the outage, and all that customer wanted to do was curse, and we listened. It turned them around, just listening and sympathizing. One of our people went out to another client the next day and the customer said, "I'm going to sue you. I hope you have good insurance." We said: "Look, we understand. We fully understand. But it was an honest mistake. It didn't have anything to do with making more money. If you leave I want you to say we were stand-up in the way we did it." The customer decided to stay with us.

Open communication is the key to this sort of unmatched service and support. Dick Padulo says:

> If you're up front in communicating, and being honest with customers, they respect you more, and you get that alliance, that partnership, where they won't go out looking for somebody else, they will come to you first. But you've got to be fair and honest. And if you make a mistake, you have to tell them, "Hey, we blew it." We caused the trouble. We had a couple of instances where we took entire circuits down, but we were upfront and honest with the customer. That kind of integrity means a lot. Today, people shop to save a few cents, that's what it is all about, saving pennies on the dollar. But I think the biggest thing we can offer, which ties into our service and our culture, is that we're not the cheapest but we're there to help you.

We began this chapter with some glimpses of how our people handle problem calls here in Rochester, yet no one is more of a master of this art than Holly Roedel Moore, in our Voorhees, New Jersey office. When we visited Holly not long ago, in our New Jersey office, asking her why she is so good at handling customers on the phone, she asked if she could call one of her customers and ask *him* to answer that question for her.

As she dialed the phone, she said:

With my team, I tell them to defuse the situation and listen. Know when to shut up, and let the customer vent. Some people want to talk over a customer. No. Just let them be heard. Keep interacting until you get to a comfort level. You can turn a bad situation into a positive one, and ultimately make them feel they made the right choice to go with PAETEC. It's a personal, human interaction.

It's all in the tone you use, how you say things with a customer. Go at the same pace as the customer. Don't talk faster. Help them understand and keep backing up what you say until they do understand. Ask for them to repeat back to you what you've explained so you know they understand. It's how you say things: it's important to use words on the same level. Don't strain your vocabulary. You have only your voice to rely on. No facial expressions. No body language. On the phone you are the single representative of PAETEC. Coddle the account. Handhold the account when it's needed. Once everything calms down, follow up, and ask if everything's OK. That's very very important.

She's waiting as the phone at the other end rings: "Ah, come on. Look at your caller ID."

Raoul answers, finally, and she says hello.

"I know this is a little off the wall, but I have someone here who wants to know if I'm any good with my phone skills. I thought you might be able to answer that," she says.

"OK. Yes. You're a little off the wall."

She laughs.

Raoul chuckles and says, "Recently, we had trouble getting some of the finer technical aspects of our service set up. I was feeling frustrated so I reached out. Holly reached back."

Holly smiles: "Well, that's nice of you to put it that way, but you were going to bomb our office I think is what you said."

"This is true. Holly lit a fire under people after a while, and they did what it took to resolve the problem. It required more than two weeks. I mean, it was two weeks once Holly got involved. The pain was around for a while before that."

"You're being nice again, Raoul."

"We were trying to build a Virtual Private Network between offices in New Jersey and New York. PAETEC provided data links and built the VPN. My account manager wasn't communicating at all and was leaving me in the dark. As problems got worse and worse, I screamed bloody murder. I emailed my supervisor, who directed PAETEC to get in touch with me right away. Within an hour and a half, there was at least one voicemail waiting for me. "Hey," I said, "I want this guy off the team." So they took the account manager off and got somebody else and that person turned out to be Holly. If you don't talk to me every two hours I start freaking out. That's what I required. I needed that level of attention."

"A little coddling," Holly says.

"You have to call me a lot. If I call and asked a question, let's say issue ABC on the table, and it has been going on a couple days, I need at least one phone call every day, and I make that pretty clear when I call. Before Holly, I could hear crickets and tumbleweed going by waiting for an update. I didn't like that. She was bringing resources to the table to get it solved. She could coordinate the people who could get it done. I may have complaints XYZ, but in the end I still have a problem to be fixed. That was first and foremost. Holly recognized that. I tend to be skeptical. I tend not to believe. I had some smoke blowing out my ears already. She communicated very thoroughly and brought resources to the table. I think we're on a good working relationship at this point."

"I call you every Friday now, right, Raoul? So, this is my Friday call," Holly says.

He laughs, and they say goodbye.

Holly explains: "It's more than just having a soothing manner. They want to get a response. Callbacks are within ten to fifteen minutes. If it's an emergency, I call back right away. Prioritize. You've got to follow up: that's the key thing. You have to reach out and keep track of what's going on. I can be his voice for him inside the company. You can't lose sight of that. I'm always prepared for the unexpected. Don't overreact. Respond only to the situation. Know the timing requirements. Get an agenda together. Anticipate their needs in advance. Get that in the first conversation. Don't overpromise. Be able to read your customer at that moment. I got to know him well, and I understood he had timing requirements, when and how often he wanted to be contacted, and I knew he liked follow-up. What are they looking for? Write down the customer's basic needs. Write. It. Down. Check it off as you accomplish it."

Holly is the personification of unmatched service and support. And other telecoms wonder why so many of our customers never think of leaving us, no matter how many problems they've encountered along the way. Caring for customers, at a personal level, is our only business. It's the heart of our caring culture. Telecom is just a pretext for the relationship. It's the people that count, stupid. Both inside and outside the company.

When you create a caring culture within the company, it naturally extends outward to customers through service and support. Employees who show how much they care about one another will behave exactly the same way toward customers. But it isn't just about business processes. Again, it's personal. I'll let the words of Brad Bono, our former COO, illustrate one of the most powerful instances of what I'm talking about, and I'll save the rest for the end of the book:

> One of our best customers in Washington, DC, Chris Peabody, helped put us on the map there by signing with us for service at Georgetown University. Not only was he a great customer, but he became a customer evangelist, spreading the word about what PAETEC could do for other potential customers. But he left Georgetown and went to work at a consulting firm, and there was nothing more he could do for us. He wasn't a customer. But we heard that his 15-month-old daughter had died. We came out in force for her funeral. I drove more than three and a half hours to get to Bethesda, Maryland. I walk in, and the line to get into the church is coming out the door. I couldn't walk three feet without running into another PAETEC person standing there. It's me and five PAETEC guys at the funeral for the child of someone who isn't our customer any longer. I see Chris shaking hands as people come through. As the PAETEC people came through he didn't shake *our* hands, he hugged us. There was no business motive in any of it, on either end. We'd become friends. We'd been friends from the beginning. It was purely because we were sad. It was personal. The PAETEC people A) came out in force and B) they *wanted* to be there. They had nothing to gain by it. I see this every day and this is different from typical companies.

Is it any wonder that, month over month, more than 99.5 percent of PAETEC customers stay with us? That's one of the best customer retention rates in the industry. Customers enjoy doing business with us because we will go to almost any length to make a customer happy, even if the problem isn't our responsibility. If a customer calls with a problem, we will stick with it until that customer is satisfied, even if the problem originates with the equipment of another vendor. When it's time to cut over service from a previous telecom, we will be there on-site at the customer's facility, to make sure it goes smoothly. But it goes far beyond all of this and reaches down to a personal level where we try to behave toward customers as if they'd been our friends for many years already, even on the first day of service. It's that friendship we consider to be the essence of our work—all the rest, the voice, data, professional services, and network expertise is just a pretext for the friendship.

PART FOUR: PERSONALIZED SOLUTIONS

CHAPTER 10
Build the Future Together

A GENUINE SENSE OF OWNERSHIP turns people into leaders, up and down the organizational chart—leadership follows naturally from the spirit. It's inevitable.

On his hospital deathbed, toward the end of 2004, drifting in and out of consciousness with his wife, Kitty, at his side, John Budney had to struggle not to lose his breath by speaking. Yet he couldn't stop talking about the business. He lay there and, as if he'd be up and around the following week, shouting into a phone somewhere, he described the kind of person he wanted to hire:

> I say to them, "I've been here since the beginning, what do you want to know?" The questions I want to hear are cultural questions. How do you take care of the customer? If it's, "I hear on the streets you are the best at customer service," that's good. I want to hear more questions about the organization and not what's in it for that person looking for a job. I'm looking for somebody who wants to get involved in something really different and distinctive. Career people. Not job people. Experience isn't necessarily important. Job knowledge has to be there, but it's really attitude and are they looking for the kind of culture we have?

A couple weeks later, John died. But before he did, he asked to be buried in a PAETEC logo shirt he often wore when he was working.

I want to give you a couple examples of the sort of person John was describing as the perfect PAETEC employee, the person whose actions are always about "taking care of the customer," which is the only kind of person you can hire if you want an organization that can come up with personalized solutions for customers.

123

Tom FitzGerald is one of those career people John was talking about. Though he's no longer with us (his work at PAETEC opened up a tremendous career opportunity at another telecom), while he was here, he was someone who walked in Budney's footsteps. He used to run our NOC and later was in charge of our vertical market operation for government clients. Tom grew up on the streets of Boston's Mission Hill district, a very tough neighborhood. All records of what he did on the streets of Boston as a teen were probably tossed out of the court system a long time ago, at whatever age those records are usually destroyed. He still talks tough, a big, in-your-face kind of guy, loud when he wants to be, but he's all heart. He dressed up as a game show host in a white tuxedo and ran an auction to raise money for the John Budney Foundation at our first Budney golf outing, and the performance he put on was worthy of an awards-show host. He has a great sense of humor. He drives a classic powder blue Cadillac Brougham with white leather upholstery. His PIN number for patching himself into company conference calls was 666.

Tom's dedication to the future of the company gave him such an extreme sense of ownership that he voluntarily took on essentially the work of three different people. He could have performed only one of those three jobs—the only one he was required to do—and still have deserved his paycheck and his bonus.

Even when Tom ran the NOC, he considered it his duty to be a salesman for PAETEC in almost any situation, attending customer events, explaining our capabilities. This is how he described attending the year's premier PAETEC customer appreciation event a couple years ago:

> I was at a function at the Rainbow Room, for all New York customers, and I didn't have one second to eat or have a drink. Our sales organization was taking me from customer to customer to potential customer to banker. All my customers were telling me, "You know what? Your NOC is the greatest." They call me directly, and they go on to name the names of my NOC technicians. I was almost embarrassed I was so proud of what I was hearing. You can feel the passion, the energy, the emotion. This is the way I feel about what I do.

Bob Moore, one of our senior vice presidents, will tell you all about the burden and privilege of having a sense of ownership over the company's future. He has to decide, on his own, which system will ultimately lead to the greatest earnings for the company three or four, or even six or seven years down the road. And, at the same time, he has to think about how we can satisfy each customer individually through our work processes and our products and services. When we asked Bob to handle our data team, I knew he would make decisions with more than his own interest in mind: he would choose what was right for the whole organization and, ultimately, for the customer. Even though we were talking about major capital expenditures for a start-up of our size, I didn't second-guess him when it came to choosing technology. I delegated all the anxieties and responsibilities to him. Here's how he recalls it:

We saw IT as a strategic enabler—a key strength for future competencies and services, not just a back office service. This is only now becoming the case, eight years since we founded the company. At first, I'm buying PCs, and it was coming out of Arunas's own money. For the first couple million bucks he was still writing the check. He'd brought some money on from elsewhere, but he was spending real cash. He said our biggest problem would be if we bought for the near term and then upgraded. The real way is to invest in the right systems for a long-term strategy. The only way you get in trouble is if you buy something and in three years you say you need to replace it. We needed to make sure we could scale up.

We had to buy hardware, Sun Microsystems and pretty big stuff. We were making a habit of buying all the crème de la crème equipment because it was the best choice for the long term. Sun was pushing storage. We had a million-dollar deal for original Sun servers. Sun said we could do this for six or seven hundred thousand if we'd use their storage. This was a big decision. We could save big money by going with Sun storage, but Kodak had told me they'd bought Sun and they wished, in retrospect, they'd gone with EMC for storage. I went to Arunas with the choice. He was at his desk, we were talking, the conversation lasted three minutes, and he says, "OK what's the recommendation. Did the team concur? Did you check the references? Then I guess we need to use EMC." I said, "But you need to understand there's a substantial price difference here." He said, "We need to build this to scale. We can't make the wrong decision now." I always talk about how a company needs executive commitment to the long term. This is a guy who doesn't know how he'll make payroll, and he says, "I trust you enough to give you an extra $300,000 for EMC."

In our first meeting, Arunas pulled out a box and had a chunk of amber with a prehistoric bug inside of it. He was using it as a symbol, trying to insure that we were aware of our surroundings. If you don't stay aware, you can be like this little bug who was sitting in pine sap and didn't know he'd be inside amber staring out from inside for a couple million years. It's about being aware of your environment.

The significance of all this is something that lasts even longer than expensive, scalable data storage: the sense of ownership that springs up when you put responsibility onto the shoulders of people who haven't been empowered that way before. They realize you respect their skills and intelligence and expect them to know how to do what's best for an organization that wants to be the industry's best in the way it solves problems for its customers. From that sense of ownership grows a natural teamwork: when everybody's an owner, everybody has the sort of emotional stake that facilitates cooperation.

And when you have that kind of emotional stake in solving problems for customers, every time you address a customer's needs you will naturally shape what you do for a customer based only on that customer's unique needs. The solutions you bring will be intensely personalized in every sense of that word: customized, yes, but also friendly, passionate, with a personal touch.

CHAPTER 11

Make Everyone a Change Agent

WHEN SANJAY HIRANANDANI CAME TO PAETEC, we knew he was good, but none of us realized what a profound impact his leadership would have on the company's future. He's a vice president of engineering, but his title doesn't capture the role he has assumed. He quickly became a change agent within the company, a champion of a new product none of us would have paid much attention to when he started telling us we needed to get involved with it: MPLS. He realized, before anyone else here did, that MPLS was one of the keys to our future by enabling us to provide it for customers, we would prepare ourselves, and them, for the future of telecom, in which signals are sent over the Internet rather than analog phone lines.

Sanjay tells the story of his leadership better than I could:

> Before I came here, I'd been involved at Cornell as part of the team of directors of networking for Cornell National Supercomputing Facility. We needed to move tremendous amounts of data from our supercomputer to others around the country. Financial firms in New York City wanted to use our data for stock and option modeling, for example. We didn't have the kind of budget to be able to buy a bunch of private lines across the country, so we had to use the Internet. Every one of our customers had Internet Protocol (IP) connectivity. Other supercomputer centers said they would help us roll out the new technologies. So our networks were test beds for MPLS.

> Security is a big element of MPLS, but it wasn't an issue for us. We weren't working in military security. The faculty was very sympathetic to the fact that we needed to play with these technologies just to see what they could do. We had professors trying to model what happens to a fruitcake that deteriorates

over time. This isn't top secret stuff. Or somebody in Illinois was playing simulated light saber battles against somebody in Ithaca. For most applications in that space you aren't that worried about security. You are trying to find an efficient way to deliver information from one computer to another.

We'd sit in the cafeteria and some would say, "It's not going to happen." Others were saying, "You take the IP network and put a tag on a packet you send around, then everybody could do a whole new set of things with it. That's how it started.

Cisco was one of the first to start introducing the technology. It was out in the research community. We knew the technology was sound. At that point I saw it coming—I saw this would be the future. So when I came to PAETEC, I kept everyone alert to this, and we built our network thinking of MPLS capability. We were one of the first companies of any reasonable size to do it nationwide, and now everybody is doing it. We had a hard time selling it at first.

The biggest test was trying it in our own network. My boss said, "I need stability." I told him I could give him stability but I needed him to push back. "I need somebody who will be hard on me. I can't sell it to Harris Beach if I can't sell it to PAETEC. I want PAETEC to use it first and hold me to the highest standards."

It was pretty flawless. We haven't made substantial changes to MPLS since we introduced it in 2002. The hardware was MPLS-capable right off the bat. Around the middle of 2002, we put in the infrastructure. By the beginning of 2003 we were starting to bring on other customers. It has started to take off. It's just going to get bigger and bigger every year.

IF THE HAT FITS, WEAR IT

The sort of initiative Sanjay has brought, as a change agent pushing for MPLS for PAETEC and its customers, represents the heart of what makes it possible for us to offer personalized solutions. The whole notion of personalized solutions has been at the heart of our success from the start, though it has become vastly more sophisticated in the last few years.

In the beginning, if you wanted your plumbing fixed while we were in there installing cables, we would have gotten out the PVC pipe. Whatever a customer wanted, we tried to provide, and that varied with each individual customer. You can't get any more personalized than that. We're still listening and responding to customers, but at a far more sophisticated level, as they ask us to help them make the transition from old technology to the new digital telecom world. Our responsiveness remains the same and the sense of ownership at an individual level remains the same. But the nature of the solutions and the way we work have changed.

If you had asked me to describe the perfect organization in the first years of our growth, I would have offered you this picture: it would be like flying in a jet where every passenger not only owns part of the aircraft, but also has a pilot's license. If the fellow in the cockpit has a heart attack, no problem. Send up the little girl from Seat 13D, find a doctor in business class, and the rest of us continue with our inflight showing of *Patriot Games*. The perfect organization would be a place where every single employee could step into the role of any other employee and do the job, at least briefly.

Impossible, of course, but this *is* the way we worked in those early years of dramatic growth at PAETEC when we numbered a few hundred. Everyone learned how to do everything, to some extent. Everyone was able, and willing, and eager, to wear any hat in the organization. Each of us shared everyone else's burden. We stepped in whenever someone else was overloaded or absent or catching a nap in a back room.

Here's how Donna Wenk describes it:

> At the start we were packed in, two or three or four to an office. Starting from nothing. I loved that role. A lot of people were willing to work but not a lot of people knew local service. They'd been in this long-distance world. Everyone was wearing a dozen hats: setting up the back office, the paperwork, creating rates, product sets. We were doing everything on Excel spreadsheets. It was fun. It was rock and roll. We were fifty people maybe. You got to do a little of everything. I started doing marketing work. Everybody was doing something they'd never done. I was meeting with website developers. I'd never developed a website. It was wonderful. Everyone was willing to do anything to help the company succeed.

What counts is how this spirit engages itself, on its own, in an emergency, as I've described in the previous chapter. On a much smaller scale, Donna remembers a particular instance of this teamwork and sense of ownership, and the passion for quality that grows out of it:

> In the beginning, the idea was that PAETEC knew how to practice The Art of Communications, which is why we have a reproduction of *The Starry*

Night by Van Gogh hanging behind our receptionist. This business is still an art for us. It's all part of the idea that telecom is a commodity, but the way you do it can be a fine art. We have the art of it down. It's not a science. It's a blank canvas out there. Everybody has access to the same ordinary tools. Artists know how to do something special with those tools, and that's what we are: artists, perfectionists. We took that to heart. One of our first handout pieces of sales collateral came back from the printer with a typo. One letter was wrong. It was an "a" that needed to be an "o." A bunch of us sat down on the floor and, with pens, we went through ten thousand brochures, stacks and stacks of them, and, by hand, corrected that one letter, on each brochure. Dozens of people sitting on the floor doing this with glasses of wine. Would we do it again? I'd do it again in a heartbeat. We cared about even the smallest detail. What's our name going to be? Where do I get paper? Go buy it at Kmart. I don't have a pen. Too bad. Bring your own pen to work. We still have that attitude.

Stock options, giving everyone a genuine financial stake in the company, is only a small way to cultivate that sense of ownership. It's about something much deeper than that, and, again, relies on all the principles in this book. You have to manage people as equals, as much as possible. When we were just starting up, we voted democratically on every system we bought in those first years. On our switching platform, we could have gone with Lucent, Nortel, or Siemens, as well as a few others. There were only about twenty of us on board at that point. We put it to a vote. Here's how Dick Ottalagana recalls it:

> If I'm running things, I say go buy me a Cadillac, and somebody goes and buys me one. We didn't do it that way. We said we'll find the money, but you tell me what I should buy. I don't care if it's a Cadillac, a Mercedes, a Peugeot, whatever. If that's what we need, we'll find the money. Tell me what switch you want. Be careful what you ask for, because you'll have to live with it. We've done that with every system. We had the right of veto, but it never came to that. We tried to empower people. Everybody talks about that. Take the turn, right or left. What if it's wrong? We'll fix it. People love it. But they're also scared. That's ownership. If you always come to me to make a decision, why do I have you here? What if it's wrong? We'll fix it.

My brother, Al felt he was taking a real risk coming on board in the early days. He'd worked hard and built a solid career elsewhere, but I persuaded him to join us, and he admits he's a little surprised at how much teamwork—the sense of ownership—he sees at PAETEC, in comparison with the kind of culture he found in other companies:

> In the early days, it was like being in a garage, more or less, bootstrapping everything. I've always had the perspective of having been an auditor and working in various different places—companies, industries and cities—and I'd never seen, in any company or business, the kind of teamwork we have here. There are natural frictions between departments. In most companies, no one likes finance because they say no. No one likes HR because they are firing you. No one likes IT because your machine crashes, and they don't show up. Or you have sales vs. back office. But that has always impressed me about PAETEC. I had never seen or experienced a place where everyone was pulling in the same direction to the degree they do here. Politics or friction between departments you always have but in those years it was non-existent. And even today it's much less than it is out there, in general. Arunas would say, "People don't know what it's like out there." I'd tell him: "Hey, you've only had three jobs. What do you know?" You would hear about it, or read about it in books, and some were like Silicon Valley where they did things differently. All the things Arunas wanted to try, but could never get done, he did here.

This culture of having a creative sense of ownership over every detail of how we do business, so that we can address customer issues as effectively as possible—this is the organizational matrix that makes it possible for a company to create solutions tailored for one customer at a time. The first step we've covered: have the right people and inspire them to be change agents on behalf of customers. Next, you need to create a *system* for listening to customers.

CHAPTER 12

Let Customers Help Lead

W HEN YOU ARE ATTEMPTING TO PERSONALIZE what you do for individual customers, you have to find as many ways as possible to listen
to them, ask them what they want from you, and look for guidance about the
course you need to take into the future. For about three years now, PAETEC
has been using an innovative method to communicate with its customers, which I've already mentioned. It's called the Customer Advisory Board.
We've organized these boards in all of our major markets, inviting representatives of our largest customers to convene several times a year in order to tell
us how to run our business. Usually I attend one of these sessions in each city
every year, or else I send other members of the senior team. Our job is simple:
to listen. You won't find many companies where the CEO will fly 2,000 miles
to spend a morning eager to listen to mid-level managers from the companies
he serves tell him what he's doing wrong. But I consider it one of the most
important roles I have.

We created this program as a way of attempting to get customers to
say bad things about us. This is a surprisingly hard thing to do. First of all,
customers appreciate being invited to participate in these half-day sessions,
usually in a conference room at a hotel owned by one of our customers or at
one of their own offices. They are impressed that we care enough to fly across
the country to sit at a table with them and not sell them anything. As a result,
they're hesitant to bite the hand that's offering them the flattery of this kind
of attention. The problem is, we want them to bite. So it's tough, when we've
filled the room with people grateful to be there, to ask, "How are we messing
up? What could we do better? How could we serve you better?"

These aren't focus groups. They aren't a soft way of selling new services under the cover of an instructional "seminar." They are opportunities for

133

customers to help us solve problems and chart a future course by becoming more sensitive about what our market wants, and they create deeper friendships with the people who depend on us to help them do their jobs, inside their companies.

"At first customers didn't understand what these were about," says one of our people who helped organize these events. "It's an opportunity for them to network with peers and tell us how we can serve them better. They think they're coming to a sales event. At the end of almost every board session, people will say, 'I was really impressed there were no sales people here. You're the only vendor who asks my opinion. I've never had a telecom vendor do that.'"

An executive who has attended many of these board meetings says: "Customers like these boards because they tell the truth. The customer lays the cards on the table. Our approach is, 'If there's something wrong, let's talk about it, and let's talk about it in front of a dozen of your peers.' It puts us on the spot. We pick people who don't have a problem speaking their minds. Most companies wouldn't select the outspoken ones. We actually seek them out because they'll tell you what everybody else is thinking."

Our first board was held in Chicago, in October 2003. We convened with representatives from ten customers. By 2005, we were doing 60 board meetings a year and bringing back a wealth of information about what our customers want from us.

After every board, we analyze what we asked our customers and whether it was done the right way. When we started, we were asking customers their most significant measures of quality and value. We'd sit in the room and go round and do introductions and put up on a white board their names, titles, what they really did (as opposed to the *sound* of their titles), how they started with us, and what their products and services were. At the end, Bruce Peters always asks, "How many have heard of a product or service you didn't know PAETEC offered?" At every board, everyone raises a hand. We follow up with the question, "Have we done enough to educate you about products and services? Do you feel you have a good understanding of our product and service portfolio?"

One of our marketing people reflects on how the boards help them satisfy customers:

> I feel so energized every time I leave a board. I've never worked for a company this committed to service. These meetings can have an immediate effect on our service to a particular customer. If I have an issue with a customer I don't call anyone, I'm empowered to just take care of it—I'm in marketing, not sales, but at some point or other nearly everyone at PAETEC is functioning as a

sales person. The CABs often thrust me into that role. I had a customer on an advisory board here in Rochester, and he had been complaining about spam. I told him we had an email scanning service. I said, "Ed you have got to use this product. I'm not going to make any money on it, but I think it will be good for your business." I called his rep when I got back. The pricing was not good for them. It was just too high. He called me, "I love the idea of this product but you have to help me, my boss will never buy it." I asked him where he needed to be. He told me. I know what our costs are. We aren't scared to share this information with customers. I didn't call anyone. I called the product manager and I said, "We have to be at this price. This is a big deal. Let's do whatever we have to do." We got it done. And this all came out of the board meeting.

A Customer Advisory Board is ultimately about building the business, but not directly and not *during the meeting*. The meeting is a way of putting PAETEC as the disposal of its customers. And when you do that, customers inevitably tell you what they want from you—and that heads you down the road toward providing a new service or product. It's exactly the way customers want services to be sold to them. First they tell us their pain—sometimes pain we're inflicting on them, without realizing it—and then we find a way to ease it. But it's the opportunity to talk about their needs, and talk honestly about something that seems as minor as the annoyance of junk mail, with one of the leaders of PAETEC—this is what they don't get anywhere else. That honest talk leads to solutions, and solutions lead to sales—but they also lead to satisfaction on both sides of the table.

What impresses customers the most, though, is that you are simply in the room with them, listening. Early in 2005, we scheduled a customer advisory board meeting for an entire morning in Orange County and some of our most important customers from Southern California were going to be there—Linksys, Westin South Coast Plaza, AIDS Healthcare Foundation and many others. I spoke briefly at the start, and told them mostly things I'm not required to tell anyone, being a private company: our revenue, our growth figures, our hopes of becoming a Fortune 500 company five years hence. If I'd had permission to speak a little longer, I might have talked about my family, but Bruce keeps a short leash on us when we attend. PAETEC doesn't do the talking. The customers do. I told them that, though we exploded with growth at the start, now we're growing carefully and steadily, like the tortoise.

"We're a nice, solid B student," I said. "If we slow it down a bit, we'll be a Fortune 500 company in five years."

In this big conference room, with a tray ceiling, taupe walls, woodwork, and thick crown molding, I spent three hours asking them how we could better serve them, and they gave me a few ideas we took back and discussed. Everyone spent quite a bit of time talking about themselves, personally: telling one another about families and jobs. Not only did it help us at PAETEC measure how we're doing in the effort to satisfy our customers, it turned this group into a small network of users, people who, from then on, would be able to communicate among themselves and exchange ideas on how to improve their telecommunications functions within their own companies.

At the end of the Orange County CAB, I handed out my business card. That evening, I invited the participants to an elegant dinner at Chanticlair in Orange County. The place has the feel of a private country club's Tudor-style clubhouse. During dinner, after some folks had a few drinks, I asked people at the table to tell a funny story about something that happened to them in the past. This made everyone a little nervous. What's the old adage? Dying is easy; comedy is hard. But somehow everyone came up with a story. Lucienne Hassler, vice president of administrative services at Sony Pictures, told a long, amusing anecdote about one of our most motivated sales reps in Irvine, Jason Perry. At the time when her story took place, Jason was on notice that he needed to improve his work or he'd be fired.

> Jason was a major account rep with PAETEC, and he had a friend at Sony who brought him in regularly as a guest to the Sony health club. So he ended up working out there for free, thanks to his friend. Eventually he was coming in whenever he wanted. The front desk just assumed he was an employee because they'd seen him so much. He'd say, "I forgot my ID," and they'd wave him through.

> So, I see him half the time I go there, and I'm watching him, because I can't figure out where he works in the company, and he comes up to me one day and says, "You aren't lifting those weights properly." I ask, "Are you a trainer?" He says, "Sure. I do training." So he instructs me, and it helps, so I listen to him. I'm still wondering all this time, who in the world he is. Finally, somebody says to me, "I don't think that guy works here. Nobody knows where he works at Sony." So I confront him, "Where do you work?" He says, "In the telecommunications area." I say, "No you don't. That's one of the departments I'm in charge of. You're busted, Jason."

> "Really? You run telecommunications for Sony Pictures? Wonderful. I'm just the guy you should be talking to," he says. "I work for the best telecommunications company in Southern California: PAETEC."

All along, he had been getting warnings about his numbers and about how he wasn't doing enough to cultivate new accounts. He'd been telling his people, "I'm going to get Sony. This is huge. I'm going to get Sony. I'm really going to get them. I'm working on a hot, hot lead." All the time he was working with me, it was obvious that I'm the person he needed to talk to about telecom, but he didn't say a thing. Now he was in a fix, though. He had to admit he was a rep, and lose his health club privileges, in order to make the sale. So he was torn between his abs and his income.

"I could throw you out of here," I say. Instead, we go to lunch, and he tells me what his company can do for me. The rest is history.

Not long after this, one of our customers at the table called another one a jerk, the way you would call an old friend a jerk, while laughing, and these people had met for the first time earlier in the evening. As a result, the so-called jerk told the story of how one of his children had been conceived. It was after brunch at Denny's. Or before. He wasn't sure. He and his wife had never determined which.

Reggie Scales, senior VP of sales for PAETEC said, "By the end of the dinner, we'd all said things to one another, and revealed things about ourselves, in ways that almost never happen among a group of people, many of whom had been strangers to one another at the start of the day. We'd become friends." A couple weeks after I got home, I got an email from the spouse of one of our customers telling me how much she enjoyed the experience and appreciated PAETEC. When I got that email, I considered the whole effort a success: regardless of whether or not we got more immediate business from anyone who attended.

What emerged from that experience, and what has emerged so often from these sessions, is that customers want a product road map from us: a clear and quick way of tapping into our full range of products and services, along with what we have planned down the road. By communicating with customers we've learned that one of the most important things they want from us is even *more* communication about what we do. As a result, we've undertaken one of our most difficult internal initiatives: to organize what we do on the foundation of a new knowledge-sharing database that will enable us to personalize our communications more effectively with individual customers. I'm going to delve into the specifics shortly, but first I want to step back further into the past.

During the years of our most dramatic growth, we responded to customer needs, in every situation, the way we respond to emergencies. In other words, when a customer said *jump*, we did an Olympic pole vault. We did whatever the customer wanted, only faster, better, and cheaper than the customer expected. That was the goal. No request was too ridiculous, regardless of whether it was in our best interest to respond to customers this way. Whatever we did was, in effect, a personalized solution. We needed to prove ourselves, get in the door, establish our footprint in the market, and then start showing customers how special we could make them feel.

We're moving to a new model for how to make customers feel that way—a consulting relationship that creates far more sophisticated personalized solutions, partnering with other vendors, devising ways for customers to keep up with the latest changes in technology. Our attitude is still a willingness to jump through hoops, but now there's a crucial difference. Sometimes the hoops are ones we create for ourselves—hoops the customer doesn't even recognize until we point them out.

The way we do business hasn't changed. Whether we're simply putting in a new T1 line or taking a year to devise a new enterprise-wide system, we listen to the customer, we find out exactly what the customer needs, and then we deliver it. Most of the time, we still let the customer show us the way. And if you let the customer give you advice, your client quickly learns to listen to the same thing from you.

Customers trust us. We quickly become the go-to person for that company. We want the customer to need our help. We want to hear the customer say, "I need you to tell me where to go." One customer we'd had for years called an officer because his fax machine wasn't working. It was telecommunications, strictly speaking. We called his administrative assistant and asked, "What is going on with the fax?"

"It's probably out of paper," she said, "I don't know why he calls you with this stuff. I think he just wants to talk with you."

That kind of trust translates into upselling. Reps can go back to a customer, who turns around and says, "I'm so impressed with the way you've handled things and gotten things done, I'm going to give you additional business."

Service quality extends up the ladder to the top. From me on down, everyone at PAETEC searches for a way to show a customer how much that customer means to the telecom. Soon after PAETEC opened for business it won a chance to prove itself to a Blue Cross affiliate on the East Coast. The customer had been relying on a competitor to install T1 lines into a specific location and a recent order had been lost or misplaced. They called us and said they needed eight T1s in seven days. Our usual turnaround was a month. Should we have said, "No, we couldn't do it?" Not PAETEC. Instead, we pulled out all the stops. Twenty people working around the clock made it happen. We got the T1s installed in six days. From that point forward, we went from being a company trying to get business to a company that had delivered results. The next time they gave us sixty days. After a hundred and fifty T1s in two years, we became the dominant player for them. In some ways it was luck, being in the right place at the right time and being able to deliver.

With employees who assume the success of the entire company is their responsibility, it's easy to require everyone to think *full circle*—about solutions, not just putting in hours. Everyone is expected to think about how every task will, or won't, solve a customer's problem. This requires porous boundaries between departments and people, with an informality and flexibility about responsibilities that many companies would find intolerable. This is crucial when you're coming up with innovative, personalized responses to customer demands.

Our main priority is fixing problems and serving customers—not guarding territory or pulling rank. On any given day, the subject matter expert could be a clerk or a VP or the CEO. The teamwork between departments runs deep. As a result, it isn't unknown for a rep who wins a vacation trip for surpassing his target to pass it on to someone in another department. In one instance, we had a sales person give up his trip, a cruise, in order to pass the reward along to an engineer who supports his switch. The focus here wasn't on who deserved the reward: it was on who created value for the customer.

WE ASK THE CUSTOMER'S ADVICE

We listen to the customer in many ways, but our most important tactic for understanding customer needs is the Customer Advisory Board. We invited *The Daily Herald* to our first CAB meeting in Chicago in 2003. Some people had cautioned against inviting non-customers to the meeting because it would appear to be a gripe session, and some others said it would be a great chance for them to see happy customers. We decided to try it. It was clear we had found an open-minded guy at the newspaper, and he was honored to have a chance to join, so we invited him. We explained to the group that Joe, the representative from the *Herald*, was not a customer, but we had decided to include industry leaders to make this a peer gathering, even if that meant including non-customers who were leaders. He introduced himself and the meeting went well. Afterward, Joe signed a five-year contract with us. He heard all the other customers talk about the quality of their service and that was enough for him. Our customers did the selling for us.

This is a theme that comes up again and again at CABs and, as a result, we've set up a new customer database that will become the platform for a far more effective effort in keeping our customers informed about what we can do for them. This may make the CAB sound like a subtle way of selling to our customers, and in the long-term it certainly helps sales, but that interpretation is exactly the opposite of what we want. Any overt selling is prohibited. We are there to listen, and we want to hear things that make us uncomfortable. As one of our best officers puts it: "We try to get them to say bad things. I always say the way to get better is to be in situations that make you uncomfortable. Tell us something bad. It's tough."

In Philadelphia, not long before our CAB meeting there, the University of Pennsylvania had been having network troubles for about three hours, and all our customers in that market—universities, hotels, hospitals and others—would be in the meeting. This made us nervous. Before the meeting, we had a lot of discussion about how we'd handle any questions about the past situation at Penn. I told our people, "Either you believe in the CAB process or you don't. We can deal with it one of two ways. You can ask open-ended questions, such as, "If you were us and you had an outage how would you prefer us to handle it?" Or you can tell what we actually did to solve the problem, and what we've done since then to prevent a recurrence. Ask the open-ended question, and the answers won't be in your box, they will be in the customer's box. Which is where we want them. We created CABs so that customers could give us advice, not the other way around."

That's what we did. Ninety percent of the suggested solutions, at that meeting, were exactly what PAETEC had considered, but the other ten percent weren't. And that ten percent was very valuable. A lot of colleges, universities, hospitals, and hotels have redundancy in their systems which kicks in. We weren't aware in many cases what the backup systems were, exactly. So we weren't prepared to help them establish ways to trigger the backup when an outage happened. We listened. And we responded. We wired our customers there directly into the NOC in order to activate problem-solving processes when an outage occurs. As a result, in some cases, we can now activate those processes directly from the NOC in Rochester, before the customer has a chance to act on it.

At our CAB in Washington, DC—one of the first ones we did after the initial meeting in Chicago—Ken Soper, our representative on the board from George Washington University, talked about the difficulty of consolidating telecom and IT departments. They wanted to merge the two—and in this they were early adopters of what we expect will be the norm, in the future. Telecom will simply be a subset of activities within IT. Ken, and a few of the other members were saying, "We're the dinosaurs and we know it."

There was a long discussion among the members about the changes digital technology will bring to telecommunications within organizations like a university. The CIO needs to learn about telecom, because it's going to become his responsibility when signals are being transmitted over the Internet, and the people running telecom need to learn how to operate in an IT environment. Everyone agreed it was a good thing to merge organizations, and they exchanged ideas on how to pace that change, and manage it.

Ken got a lot out of that. In that first meeting, the group covered how the two departments would interface. We offered as much insight as we could, at that point, on how the changes would help increase an organization's competitive edge. We focused on their stories and said almost nothing about PAETEC. Three months later, in the second CAB session, we described how we could help them manage the shift. In a third session, we brought in one of our IT people, who gave them, essentially, a case study on PAETEC itself and how we've leveraged data technologies. One member talked about how his company used AT&T to handle all its data but he used PAETEC for voice. He compared our PAETEC Online customer portal to AT&T's version, and ours came up short, in many ways. We've since begun to expand the functionality of PAETEC Online, partly as a result. The meeting became, for awhile, a forum for this customer to sell others on the advantages of AT&T! We didn't intervene: the CAB was assembled to help customers find solutions to their problems, and they were doing this.

When he was finished, all the positive talk about a competitor didn't hurt us at all. Ken renewed his contract for voice service with us, adding some additional 800-features. The others went away knowing we would expand our PAETEC Online to do some of the things other carriers were offering through its portal. It's all about listening and doing what customers want based on what they tell you.

When we held a CAB meeting in Rochester, we invited people from the NOC to attend in order to ask our customers how they rated the service they got from the NOC when they had a problem. We explained how the NOC was managed, and how, internally, we measure its effectiveness.

Bruce Peters asked the group: "On a scale of one to ten, with ten being the highest, how would you rate our NOC?"

The group ranked the NOC at a six or a seven. Internal metrics had ranked the NOC at a ten.

"Explain to me why it's not a ten for you," Peters asked.

As members suggested improvements, Peters wrote them down on a white board: changes they believed would enable them to rank the NOC closer to a ten. Of course, these were all things important to customers, not NOC managers. One of the issues, which has become something we continue to stress, was that, if you're a customer calling in, you don't want to hear any suggestion that the problem is not PAETEC's responsibility. It can be hard for the NOC technician who knows the outage originates with another carrier's service or another vendor's equipment, to keep on taking responsibility, especially when the customer is justifiably irate. It became very clear to us that, from the customer's point of view, the person who answers the phone can never intimate that it's not PAETEC's product causing the problem. The customer doesn't care. The customer pays us to provide service, and when the service isn't there, he or she wants PAETEC to take ownership of it. It's human nature, under stress, to say, "It's not our fault," especially when this is completely true. But we learned that we can never suggest that. Later on during the debrief after the problem is resolved, once the customer is reasonably happy, we can explain how the problem originated.

Overall, from our experiences with CABs, we've learned many extremely useful things about what our customers face and how they view us. They want to be approached with collaborative, consultative solutions to their challenges. They want to know more about what we can do for them. They will introduce us to their data decision makers. They don't want to be cold-called.

In the past couple years, we've also discovered that our customers:

- Want to provide higher levels of service to their organizations with fewer resources and declining budgets.
- Are very concerned about VoIP and how they need to adapt to it.
- Want expertise in establishing network security.
- Anticipate converging voice and data departments into one group reporting to a CIO or CTO.

We are taking all of this into account as we develop our own business strategy for the next five years: primarily in our initiative to move swiftly into next-generation technologies such as MPLS and VoIP.

CABs are only one way we've organized to listen and respond to customer needs. On a one-by-one basis, we now are trying to come up with special solutions to large-scale customer challenges. PINNACLE, our proprietary enterprise-wide software platform, gives customers much greater control over their internal voice and data network, and has become one of our fastest-growing sources of revenue. It's a product that allows PAETEC to partner with a customer to create personalized solutions to the challenges of managing an internal network more effectively and less expensively. At a user's group meeting for PINNACLE customers in the Northeast, Dan Wilson of Binghamton University reflected on how PINNACLE has enabled him to find solutions to his own problems:

> PAETEC has responded to our input in a couple key ways. They've moved the PINNACLE technology from being subscriber-centric to service-centric. And they've created a pre-order staging area where we can work out the details of an order without making it permanent in our record. Previously, as soon as you generated a work order, there was no way to remove it from a subscriber record. The pre-order is like a sandbox where you can work out the details before you enter it into the permanent record. It's a big help.
>
> People pay for telephone, cell phone, pager, and data, and PINNACLE enables us to track all of these services individually, for each subscriber at the university. In the past, we couldn't sort data by the type of service. We couldn't even sort by subscriber. You would have three different records, in three different places, for one subscriber. One for his or her telephone service. One for cell phone. And so on. You couldn't even quickly say that a person had three or four different services. You would have to go to three or four different records to discover that. Now we have access to all the data we need at our fingertips, and from that information we can manage the whole system much more effectively.

This is a far more sophisticated approach to our business that we had in our earliest years, when we installed a T1 line, turned it up, and answered the phone if the customer had problems with it.

We discovered PINNACLE when we merged with CampusLink. In the process, we brought Bob Schwartz, the CampusLink president, into our organization, and he's been an asset for us ever since—and a one-man personalization service. He was doing, back then, what PAETEC will be doing more and more in the coming years: creating technology service solutions for clients. His role for PAETEC is unique. He works with the freedom and independence—and an entirely commission-based fee structure—of a sales agent, but he does most of it in collaboration with our direct sales team. His most important role has been in helping us leverage the PINNACLE software product as a key element in our solutions selling approach.

I have a position as master agent but that's just how I get paid. Actually, I work with direct sales teams and get plugged into the profit model as a referral agent so I can get paid. Every commission check I get has to be approved by the sales organization. We sign the account and, based on my role, we agree on what my percentage will be. My contract says the actual percentage will be determined deal by deal.

PINNACLE was a third-party platform we used at CampusLink to help clients do their billing. After PAETEC bought CampusLink, I pushed for us to acquire PINNACLE. Conversations started going, and it worked out. It's a big product. It's a great product. The biggest issue is how to integrate it, leveraging PINNACLE to be a driver of other core PAETEC products. It positions PAETEC as a consultant: that's my history as a systems integrator consultant. I've been involved in the business development group at the high end, working with our partners to bring solutions sales to their customers. I don't fit the profile of a classic agent who influences a lot of people's carrier decision-making. Often, to make the right decision, you need some network in place, and that's where I come in. PINNACLE helps drive sales within PAETEC but the key thing is the way it gives customers a new level of control over their own networks.

PINNACLE is opening doors for us everywhere. Lou Surman works on strategic market accounts for us in Florida. The situation we face, and the unique way in which PAETEC can position itself now, is especially pronounced in Lou Surman's account of how we managed to get ourselves into the bid to provide telecom for all of Miami Dade County in Florida.

We won the account with an online auction. The market is so price-oriented for carrier service now, these online auctions are more and more common. It's a win based solely on price. We won on 1.8 cents a minute. At that rate, we're saving Miami Dade a million dollars a year. This is a customer with

837 locations and more than 100,000 phone numbers. But we almost didn't get into the auction.

First of all, we made sure how low we could go and still make a narrow profit. We ran the Profit Assistant software on the account beforehand. It told us how far down we could go. But we never would have been allowed to bid if we hadn't told the customer about PINNACLE. They weren't interested in talking to us because they had to go out to bid. I had one thing left to talk about: PINNACLE. We were on our way out the door, and I mentioned PINNACLE, and they wanted to hear more. We had a four-hour lunch. Eight months later, we did a $300,000 deal with PINNACLE. Miami-Dade was capturing only 10 percent of its call records—it couldn't pinpoint where their costs were originating in 90 percent of their call volume. Now, with PINNACLE, the county can track all 100,000 numbers. It also allows Miami to be 911 compliant: if a call goes out anywhere in the system to 911, it can track exactly what phone is being used in which building. We essentially did a study and came back with a white paper and said here are the inefficiencies and here's the value you'll enjoy with PINNACLE. PINNACLE opens doors. There's nothing like it.

Jim Lancer, a product consultant with our PINNACLE professional services support group, describes why PINNACLE offers such a unique benefit to customers:

I've been dealing with large strategic customers, mostly universities.

A lot of times in a package, one element does one thing and another element does another, but they don't talk to one another. What we're offering is an end-to-end solution. You get billed from your vendor and PINNACLE helps you manage that cost, but it also helps you manage internal operations where you're creating services to an internal environment, and it helps you bill your internal customers for a service. What this means is that our product can help you match what you pay with what you bill out. You're able to pass out costs to the profit centers in your organization that are generating those costs.

But that's only half of the application. It has two modules: the cost manager and the operations manager. The op manager helps you do service order requests, for the cabling and infrastructure and switch programming and features of your network. It tracks inventory, what kind of phones you have on desks and what you stock in your warehouse. It tracks technician time, and helps you schedule it.

Customers love it. They come to us for PINNACLE and then end up asking us to be their carrier.

Jim came to us essentially right out of college. But he came to work for us after being a customer. Within months of his graduation, he took a job with his college's IT department. In that position, working for the college, he bought PINNACLE from us and implemented it there.

I worked at Keuka College. I graduated in May of 2000. I was a student and graduated and then worked as an employee. I worked in the IT office as a system administrator. I bought PINNACLE in July of 2000. My boss helped me implement the software there. In October, I took a job with PAETEC.

It's how you set up business rules that make it unique. No two customers use it the same way. They implement it differently by turning on different kinds of parameters. A big selling point when we come in from professional services is that we're here to help them create solutions. That's why when people purchase software they make large investments in professional services, because we provide technical expertise but also we show customers how to implement best practices based on what other customers are doing with the product.

I think the classic example of setting the bar high is the invoice manager, which gives you an engine to analyze how your vendor invoices you. We give a customer PINNACLE, and we tell them to use it to check how well we're billing them. If PAETEC is billing you for 50 T1 lines, and PINNACLE says you only have 40 T1s, then there's no way around it. We'll bill you for the 40. It's a confidence factor. It's a good way to audit whether you are being billed correctly. Our stance is that if we aren't billing you properly, we want to correct it. With PINNACLE, we're installing a watchdog on our own services.

When we assigned Jim to work with Stanford University, he spent two years virtually living on campus helping them reorganize their network services. They were so pleased with his work that when he was finished they came to us and offered to write a bonus check for $20,000, which they wanted us to give to Jim. This was a sum of money, unsolicited by us, above and beyond what they had contracted to pay for Jim's work. That's how valuable he'd been to them.

Jamie McDowell, a senior account manager for PINNACLE, travels around the country, from Alaska to Florida, calling on existing accounts to explore how we can improve what we do for them with PINNACLE, explaining new products and services we offer. It's part of how we keep the business as personal as possible: as I mentioned before, we possess the ball for more minutes during the game than the competition. With PAETEC you see the face of someone who works here more often than with almost any other telecom.

> Many customers purchased the product in the late '90s, and at that time they bought it to charge back costs to departments, which is still a huge benefit with PINNACLE. They also used it to manage resale of telecom services to students, which has declined with the rise of cell phones on campus. Charging back telephone charges to departments is still big, and now we can charge back anything IT handles.

> PINNACLE isn't a commodity. We don't live and die by fractions of pennies per minute as the average telecom rep does. It's more than a product: it's a new way of understanding your own telecom network. We call it a business process improvement. A customer buys a package initially that's predominantly labor. Our people go in and look at how they are doing things today and whether or not they should be doing certain things with PINNACLE.

Sony Pictures Entertainment (SPE) is a great example of a highly personalized PAETEC solution. They were using multiple carriers. We started, as usual, by replacing three T1 lines, our bread-and-butter business. The relationship grew and we introduced them to PINNACLE.

SPE supports an extensive telecommunications network to serve 10 sites nationwide, including 1,100 departments and 200 production companies. Distributing paper bills to each department and production company was both costly and time consuming. As incredible as this sounds—like something out of the movie *Brazil*—the company employed 150 people simply to print and distribute phone bills to staff in 1,100 departments and 200 associated production companies. These bills would go out daily, weekly, or monthly, depending on the project. If a production company was filming a commercial in one studio, it would get a daily phone bill for calls from that studio. If it was producing a movie, the bills would arrive weekly. PINNACLE software enabled Lucienne Hassler and her team to automate much of this process, generating electronic bills at the touch of a key, with detailed usage reports available for any phone, user, facility, or production company.

Customers can even log onto the web and examine their bills electronically.

By doing all of this, SPE has saved $250,000 in labor, paper, and distribution costs. Revenues to Lucienne's department, from telecommunications, have risen by 50 percent. Her previous system often crashed, allowing many of the 33,000 daily calls to slip through the cracks. Since PINNACLE has never gone down for SPE, the company hasn't missed a single call. In addition to voice services, SPE can now bill for a variety of other activities including DSL, teleconferencing, wireless, fax rentals, pager services, and phone equipment. And SPE can run extensive reports that track charges for every item in each department and company it monitors.

"The reliability of PINNACLE is of paramount importance in providing our customers with accurate and timely bills and ensuring that Sony Pictures Entertainment receives the revenue it is due," Lucienne says.

This sort of solution is a model for how PAETEC intends, eventually, to sell to all its customers: establish a consulting partnership, analyze work processes and the unique challenges of a particular business, and then create a customized solution for that business. What we have done for Linksys, a division of Cisco, is probably the most outstanding example of how we create solutions.

CONSULTING FOR PERSONALIZED SOLUTIONS

Linksys is PAETEC's largest customer and one of the best examples we have of a large-scale personalized solution. The company pays us almost a million dollars per year for our services—PAETEC handles a large portion of its telecom service. We enable Linksys to manage the massive amount of toll-free traffic it generates—the incoming help desk calls from people who buy products, mostly routers, from the company—in a way that creates customer satisfaction and ultimately lowers costs.

This partnership began several years ago, when Linksys chose PAETEC for its local and long distance telephone service. Before Linksys was acquired by Cisco, Linksys came to PAETEC and made a bold request. Demand for the company's new wireless router was exploding. Linksys had always been a market leader, but now it was a star, and with the spike in sales came a rising tide of help desk calls. Linksys wasn't equipped to handle the new traffic: people were being put on hold, sometimes, for hours. The company needed a better way to handle international toll-free calls from customers who needed advice or service. Linksys requested a new software product, one that would automatically route incoming calls to one of its various help desks around the world—software that would send calls to the location where customers would get the quickest service. Linksys asked for this product—which didn't exist at the time—in October, 2002.

The problem for Linksys was that the "false busy signals" its system would generate to callers were a burden on the system. It wanted, if nothing else, a way to offload those signals onto the PAETEC system, where they could then be routed to help desk staffers when they became available. PAETEC knew it could do more than that: it could bring more intelligence to the routing process itself, monitoring in real time who was available to take a call, around the world, and who wasn't.

More than a few people at PAETEC saw disturbing implications in this. The company was making money on every minute of those toll-free calls. The longer people were put on hold, waiting for help, the more money PAETEC could make. Was it in PAETEC's best interest to come up with a solution for Linksys that would dramatically lower the revenue PAETEC could make from Linksys? There was a tense meeting late in 2002, when three key people met at PAETEC's regional office in Irvine, California, to debate the merits of giving Linksys what it wanted. The three were Gary Eisenberger, Mark Szotkowski, VP of direct sales in Irvine, and Peter Orosco, director of engineering in Irvine.

"Here's the thing," Peter said. "When we first implement this service, the amount of time it will take the company to help customers will drop nearly in half. That's a huge hit on revenue up front. Those lines just won't be as busy."

"I'm not sure we should go through with this," Gary said. "We're going to cannibalize ourselves with this. It will cut into revenue."

"That's the initial effect. I agree. But long term, it will increase revenue. Trust me," Peter said. "Once we do this, once we minimize the hold time for people, the volume of calls in any given period of time will increase. Linksys is on a growth curve. The company will be able to do more in less time. Customers will get much faster service. There will be fewer unanswered calls. And we'll end up making more money. It's a win-win-win. The average hold time is going to drop. That will improve productivity. And as their business continues to grow, so will ours."

Peter believed it would increase revenues, not cut into them, but the key thing was that this is what the customer wanted and it was good for the customer. It was a long-term consideration: whatever, if anything, PAETEC lost in revenue, it would gain in customer loyalty and referrals. Peter pressed hard for the project.

This is where the culture of PAETEC does justice for our customers. Every other carrier was layered with bureaucracy. A request like this would have had to go through many layers before anybody would even say, "Yes, that's a great idea," and then dismiss it as impractical.

When we came to this crossroads with Linksys, it was losing its ability to manage its call traffic successfully. On top of that, *PC Magazine* had just printed an article saying Linksys was rated one of the worst call centers for support. Victor Tsao, the CEO, was not happy. Everybody was under the gun to find outside resources. Customers were being put on hold for three hours at a time.

The development team, eager to solve the Linksys crisis—not sure whether or not it would be good for PAETEC, but loving the challenge—got the green light in October of 2002. In only four weeks, the team had a test model of a new software product that would monitor call traffic at all of the Linksys help desk locations around the world and route incoming calls to any technician who could respond quickly to the call. After two months of trials, on Jan. 15, 2003 the product was ready for prime time: Toll-Free Busy Overflow. PAETEC now offers it for any customer who needs it.

When the system was up and running, sure enough, usage of the toll free lines dropped dramatically, and so did PAETEC revenue. But within three months, with the continuing growth in business at Linksys—made easier by its new ability to help its customers quickly and effectively—the usage of the toll free lines rose above its previous levels. Peter Orosco was a hero. The average time customers were kept on hold dropped from seventeen minutes to nine minutes.

We had shortened the amount of time Linksys worked with customers by simply cutting out the dead time when they were on hold waiting for help. Linksys could, effectively, serve twice as many people in the same amount of time. We were in constant contact with the customer as everyone working there became accustomed to the new system. Day and night, we made phone calls, paid visits in person. We were at everyone's fingertips. One day, Linksys called and asked Peter to come in. He went to the office, and he could tell it was buzzing with some kind of good news. Ross, his customer, greeted him and took him to the back of the plant, into a big room where there was a huge banner: *Best Rated Call Center in the World*. The company had just been awarded this rank from *PC Magazine*. Everyone was ecstatic. The Linksys people had organized this surprise party for Peter's customer, and Ross had called him because he wanted Peter to be there to share in the glory.

But that was only the beginning.

A year later, Cisco acquired Linksys, and, again, fear rippled through the PAETEC ranks. Many were certain Cisco would switch Linksys over to its own supplier for telecom and data. The mood was grim. Some were certain that other carriers would do anything to have the Linksys brand on their list of customers. Everyone was waiting for the phone call that would bid PAETEC farewell. In fact, PAETEC later learned that Cisco's largest customer, who happened to be another telecom, offered to buy out the Linksys contract with PAETEC for $10 million. The competitive pressure was intense.

Sure enough, the call came in. They asked for a meeting. I went down with several others from my team, and we met with their team. The new Cisco leaders told us that Linksys could have anybody it wanted for its business. Whatever was good for its bottom line. Yet they said, "Our team here at Linksys tells us they have a long, long history of great support and service with you. So we're going to stick with PAETEC."

PAETEC signed a deal for three more years with the plan to greatly expand its service to Linksys. The client wanted PAETEC to completely take over the call center operation. It wanted to essentially outsource all the call routing for the help desks to PAETEC, hardware, software, staffing, everything. At the same time, it would move to a more sophisticated level of IP routing. Together, we have developed a technology and process that allows Linksys to have all call centers communicate in real time to a central controller that has all the information about who's on line and ready to help. For each call, PAETEC will send a query to a central controller: where should I send this call? It will look at every call center specialist and discover someone in, say, Manila, who can take it. In other words, PAETEC's system serves as an intelligent central switch for all these different businesses—doing essentially what, in the beginning of telecommunications, back in the day of Bell Telephone, a human operator did on a much smaller scale.

Beyond calculating simple availability, the system will also analyze the nature of the call, the technical details of the problem—even the language capabilities of the caller—and match the caller with the help desk technician who has just the right skill set. It's a sort of matchmaking software, on top of the former function of simply routing a call based on the workload of the various help desks. This is a "third generation" application: not many carriers have it, but any carrier can buy this kind of equipment. PAETEC would be the only company at the time outside the big three—MCI, AT&T, and Sprint—that would have this system.

Over the Thanksgiving holiday in 2004, Linksys implemented large changes to its system in the process of shifting over the call center routing entirely to PAETEC, and these changes occurred four times a day. Our engineering people on the East and West Coasts worked through the holiday, implementing the necessary changes.

They had to go way above and beyond their usual work load, with labor-intensive manual changes to the system. We were going into Thanksgiving weekend, but we stuck with it and made the changes. At three o'clock in the morning on Thanksgiving Day, Peter's cell phone started to ring. He picked it up and it was Marion Wyand in Rochester. "Peter! Peter! The changes we've made will take Linksys down." She had been thinking about the changeover in the middle of the night, and it occurred to her that the changes might have a negative impact on the system. She got up at 3 a.m. to make the necessary adjustments to the process before it had an impact. By 6 a.m., she had been up three hours rounding up troops to get the changes made to get everything done for our client.

At this new level of service, PAETEC is actually learning from Cisco. The parent company's call manager is far more robust than the technology PAETEC developed, and Cisco is essentially giving it to PAETEC to use as a tool for helping Linksys. In this partnership, Cisco and PAETEC worked for nine months and prepared to move this system off of Cisco's platform and onto PAETEC's.

And yet, Cisco and Linksys believe they will continue to rely on PAETEC for new technical insights, new ideas for solutions, and possible improvements for a new generation of technology. As a result of this partnership, we now have a new service called IP Contact Center, in which we offer other customers the same kind of solution we built for Linksys. What was a one-off personalized solution will have become standardized enough to sell as a product that can then be personalized again for another customer.

We've talked to Cisco and Linksys about creating a technical consulting team or technical advisory board team, because we want to expand our partnership with them. We like the idea of presenting solutions to them on how they can operate more effectively, once we get the system up and running under our roof.

We can cite hundreds of examples like this, although none are as sophisticated and complex as the solution for Linksys and Cisco. Our customers are beginning to realize how unusual it is to have a telecom provider that brings to the table this kind of engineering capability combined with this level of caring and resolve to come up with solutions to challenges posed by advances in technology and the need to cut costs.

Looking Back and Looking Forward

I CAN'T THINK OF A BETTER WAY to look back on what PAETEC has accomplished so far than to have existing customers reflect on why they chose PAETEC and why they are loyal to us. It all comes down to the four principles: caring culture, open communication, unmatched service and support, and personalized solutions. When we call on a new customer, we can offer the names of dozens of customers who will describe what makes us different. In each and every case, they will confirm that PAETEC simply goes further than any other telecom they've dealt with in our efforts to make them happy.

A few parting examples:

CARING CULTURE: THE CITY OF CONCORD, MASSACHUSETTS

"I feel more like an individual, someone PAETEC cares about. PAETEC service is outstanding. A year ago, when AT&T was working our phone systems, it sent out a message that it was going to cut the lines because it had some work to do. I had a computer system up so I tried to get some work done that day and the next day was supposed to be the day that the power was going to be cut off—but AT&T ended up cutting off service THAT DAY, so I lost all connectivity. That's why I feel very comfortable with them, because I know where PAETEC stands and what it is going to do for us. It provides us with the type of service we have never experienced before. It's a refreshing thing. We have a comfort level with PAETEC people; they made us feel as though we are the ONLY customer they have. This helps us understand that the company is up there not to just make a quick dollar but to be able to provide a service that supports the needs of the customers. And the bottom line is: by switching to PAETEC, we save $36,000 a year, and that will increase as long distance rates drop.

"When we were switching from Global Crossing to PAETEC, at first the decision was overruled because we were concerned that PAETEC was a private company and the telecommunication industry was in trouble, so we were directed to AT&T—it was a known and trusted brand. But ultimately we turned to PAETEC for service.

"The service is exactly the way we like it, never a billing problem, its just been terrific. We've used Sprint, and we switched to Global Crossing and Adelphia. But with PAETEC we can integrate long-distance voice dialing for all of our regional vice presidents, and get all of our activity on the same bill. It's fantastic. Response is wonderful. I call with a problem, and boy, I don't have to do much else but call one of those cell phone numbers—the skids get greased right at that point! Words to describe PAETEC: partner, attentive, responsive, cost-effective.

"Whenever we call, the person on the other end actually gives a damn about what we need. The person on the other end really tries hard to fix it. In fact I'm hard-pressed to think of any difficulty we've had in getting something resolved. If we need anything, we get on the phones and we get action. It's great."

OPEN COMMUNICATION: KEYSTONE MERCY HEALTHCARE

Although institutions may be hesitant to migrate voice services from a large, national provider to a small communications company, Keystone Mercy Healthcare, in Philadelphia, took the first step when they contracted with PAETEC for a toll-free origination several years ago. According to Keystone Mercy Healthplan's Senior Project Manager of the Voice Network Service Group, Beth Seymour, "We had been looking to replace our local service provider, Verizon, for quite some time. I thought the cutover to PAETEC service was great—it was seamless and invisible to the end user." Despite two instances of temporary equipment failure, "PAETEC was attentive and responsive" and implemented additional redundancy to prevent the issue from happening again. In Seymour's case, her PAETEC account manager handles everything. "I've never had to call and alert him about any of my problems, because he always knows what is going on." Not only is Seymour able to contact several individuals when necessary, but all associated parties constantly work together, communicating with one another on the same issue. "My experience with the NOC has been very positive," she continues, "its so nice to call and speak to a human being!" She did not get this same type of service at her other provider.

Keystone Mercy Healthcare was more than happy to re-sign the contract to stay with PAETEC for 3 more years. "And why not? We always feel like you have us in mind when we call in. We feel comfortable digging our toes in deep with a company like PAETEC. It has provided us with the best service at the most cost-efficient price, I have never had a reason to look anywhere else. It's a trust thing, why we stay with PAETEC. Its critical. Our lifeline is our phone. We trust PAETEC to keep us in business."

PERSONALIZED SOLUTIONS: GRAND SUMMIT HOTEL

Mark Giangiulo, general manager, recalls: "A random man was calling the hotel and making terrorist threats to the guests. The man had not only been doing it at this hotel, but all over the country for many years. We contacted PAETEC people about this problem, and they went out of their way to trace the call and find out who this person was. Within 24 hours, this person was apprehended, and is now behind bars. He would randomly pick up the phone book and make threatening calls to hotels, apartment complexes. He would describe things that they should be doing to prevent them from being killed, and he told them that he was watching them, the whole bit. They found him, and they arrested him. PAETEC actually traced out the numbers and cooperated with the police to apprehend this terrorist. Typically, you call and the providers will laugh at you and say they are not going to help you. He was actually arrested under the Patriot Act, for making terrorist threats, a federal offense."

According to Giangiulo, Grand Summit's cost savings have been tremendous, but the most significant benefit has been the reliability of the PAETEC network. "Customers are very demanding," stated Giangiulo, "so our biggest concern is that the system has to work. The most important thing is that the customer is always able to call, always able to connect. And PAETEC has delivered that reliable service."

I have to admit, with customers like these, it's easier to look back than to look forward. When you succeed at something, it all seems so clear in retrospect, as if you knew, every step of the way, what was going to happen next and how to handle it. In reality, we discovered much of what we had to do simply by doing it. And the road ahead is going to be appreciably more difficult, but also more exciting. The real tests for PAETEC are still to come: how we manage to grow through the next five years will refine us into becoming an even more unique force in telecom.

Our market is changing in dramatic ways, and PAETEC is currently racing to adapt to those changes. In only eight years, we've recapitulated a quarter century of change and adaptation in corporations like Eastman Kodak Co. and Xerox Corp., as they retrofitted and reengineered to serve a market transformed by digital technology. In our first four years we had enormous growth simply entering the traditional local and long-distance voice market. We opened the tap and the water flowed. In just a few years, though, this service has rapidly become a commodity and prices are dropping, never to rise again, as the Internet offers unlimited voice services as a simple add-on to data, and as cell phones steadily erode traditional land-line revenue. We are having to reposition ourselves less as a provider of something everyone else can offer and, instead, form long-standing consulting relationships with customers to offer solutions to complex voice and data problems.

The challenge we face over the next five years is to change and improve the way we do everything—here's the hard part—without straying from the four principles of this book. Two key challenges require us to re-engineer. Cellular and Internet technology is eroding what has been our core business until now, traditional local and long-distance service. Second, we've grown to the point where some of the internal processes that worked for us in the past just don't do the job anymore. We have to become more systematic and sophisticated in the way we do everything.

This means we're changing our approach to our market, in the near future, by focusing on making our existing customers even happier, hoping they'll consider us worthy of a larger share of their telecom budget. In 2006 we're going to dance with the ones who brought us here. Our eyes will be primarily on these current partners, rather than looking around the room for somebody new.

When we were gearing up for an IPO, which we were doing in varying degrees throughout the first eight years of operation, we were obsessed with one number: our earnings before interest, taxes, depreciation, and amortization (EBITDA). We worshipped that number. We cut costs to strengthen it. We held the line on investment with an eye on it. We did nothing to lower that annual number and everything to raise it as much as possible. But, as I've pointed out, every time we were ready to go public—though we *could* have—I refused to pull the trigger because it would not have been in the best interests of our people and shareholders. So we've put the IPO on hold, and we've become obsessed, instead, with serving our customers even more effectively than ever before. We're improving what we do for our customer base—which is our way of trying to establish new ways to live up to our four core principles, ways which will see us through the next five years.

As a token of that initiative, at the time of this writing, here are a few things we're doing.

We're adding more than a hundred people to our payroll in 2006, virtually all of them brought on board to upgrade existing processes or create new ones to increase customer satisfaction. This hiring will help us improve the way our existing departments do their jobs: customer service, operations, IT, and engineering. But part of that improvement will be around processes that enable our departments to work together to personalize what we do for customers. We have systems that still don't communicate as effectively as possible with one another. Our Network Operations Center, at the time of this writing, still can't tap into our store of knowledge about our customers. It will soon. But this is the sort of challenge we've never faced before, because we've never had such powerful tools to help us make things even more personal. Soon, when a customer who has network trouble is working with a NOC technician to solve the problem, that PAETEC employee will, with a few keystrokes, be able to call up everything we know about the customer. It will be like talking with a long-standing friend.

In our compensation policies, we're making changes, as well, to encourage better customer service—we're targeting financial incentives to strengthen account development by sharing ongoing commissions with the sales organization for existing customers. If we think about the customer first, we'll be rewarded with a larger share of that customer's telecom and IT budgets. That's how we hope to grow: by deepening our existing relationships.

In our emerging new line of products, we're racing into the new world of Internet telephony and professional services. Cisco Systems has told us our VoIP offering is one of the best in the world. This is an endorsement from one of the world's leading network technology companies. In addition to VoIP, we're offering IP Contact Center, the personalized call center system we developed for Linksys. We're offering, as well, Managed Security Services, Hosted Firewall Services, Email Scanning Services, SECURETEC MPLS, and TRADE TEC. All of these emerging PAETEC products draw us directly into a world that requires a whole new way of operating. They also oblige us to deliver highly personalized solutions to our customers who are increasingly found in Information Technology rather than telecom departments. In all our early years of growth our revenue could easily be estimated by the number of T1 lines we put in for our customers. Now we're looking toward IP services as the major generator of new revenue. The challenge is this: though we may have great IP products, in the surge of demand we're experiencing, we need to rapidly scale up our systems simply to process orders, initiate new service, and bill customers for it.

Many of the changes we'll be implementing in the next few years are entirely customer-focused on improving the way we communicate with our accounts. We've been redesigning our private web portal for customers, formerly WebFront and now PAETEC Online. It's our way of opening ourselves up, so that a customer can sign in and manage an account immediately and directly. Though only ten percent of our customers take advantage of this convenience, we want that number to jump dramatically. We've benchmarked PAETEC Online not only against other telecoms, but against user-focused websites like Amazon.com, and we're satisfied that our customers will find it as friendly and helpful in personalizing their relationship with PAETEC as anything else out there.

Until now we have never formally measured customer satisfaction, but we're preparing to implement a program to measure it directly. We serve around 17,000 business customers. We plan to be able to measure how each and every one of them feels about our service by the end of 2006. Personally, over the next year, I intend to meet or speak directly with our top customers around the country, meaning anyone who bills more than $10,000 a month with us. At some point, I will be face-to-face with people from each of those companies, asking them how I can do my job better.

With all these changes underway, our already outstanding internal training program is growing to help prepare everyone in the company with the knowledge they need to keep up with all these changes. Here's an admission that, quite frankly, both encourages and frightens me. At the moment, we don't employ many people who could sell everything we offer—data services, PINNACLE, hardware, and the original local and long-distance services. This is good news because it demonstrates how dramatically we've grown. But it's scary. We're scrambling to close that knowledge gap and we'll keep scrambling in the years to come. Still, the core of our business is familiar to all of us. Our revenue from long-distance in 2005 was $173 million; from local voice traffic, $219 million; and from data services and products, $62 million. There's no knowledge gap in the first two segments. Yet the gap is huge in the third segment which is by far the fastest-growing one for us. It's where the future is hurtling towards all of us in telecom.

EPILOGUE
The More We Change,
The More We'll Stay the Same

THAT'S MY VERSION OF the familiar old French maxim—*plus ça change, plus c'est la même chose.* Essentially, it means the more things change, the more they stay the same. What it means, in my version, though, is something a bit different. As we've seen, there are plenty of new things under the sun in telecom. But when I say the more we change, the more we'll stay the same, it isn't so much an observation as a dream and, therefore, a promise. It's a promise we are making to ourselves, our families, our shareholders, and our customers: no matter how much we need to change to adapt to how the market is driving innovation at PAETEC, we will keep looking to our principles as the four points of our compass. It represents our passion for quality through caring culture, open communication, unmatched service and support, and personalized solutions.

A giant step, and a major milestone, on our journey took place in June 2006, when we converted all of our preferred shares to common shares. In doing this, we bought out the holdings of a few major stakeholders whose ability to influence our future had become too burdensome for our vision. We believed the privileges that came with their preferred position would have given them far too large a percentage of the company's value, were we to go public.

The capital markets made all of this possible. As various telecommunications companies merged over the past two years, making bankruptcies a thing of the past, and eliminating much competition for the surviving companies, more debt became available to us. We looked like a much better bet to those willing to loan us enough money to pay off those investors. We effectively bought back 60 percent of the company and put ownership squarely into the hands of our employees, who now hold a majority of the shares. I am

159

a common shareholder, along with everyone else now. As part of this move, our board of directors has been reduced to key people who share my vision, from inside and outside the company.

So, the leveraged buyout of some of our large key investors opened up, for us, a new path for growth. Our cash situation not only enabled us to free ourselves from foot-dragging on decisions such as when to go public, it gave us the ability to expand through acquisitions. And with all shares converted to common stock, and with majority ownership of the company in the hands of employees, we were now free for the first time, to make a strategic acquisition swiftly and decisively. And we did, almost immediately.

With our new streamlined team, we will be able to merge with another telecom in record time. We will merge with US LEC, in a negotiation that took only six weeks to resolve itself into an agreement. As a result, we should become a $1 billion company by early 2007 with 45,000 customers. The merger with the Charlotte-based telecom will greatly expand our presence throughout the Eastern United States, where our customers have been hoping we'd expand. It gives us a new foundation, which we can use as a springboard to become a Fortune 500 company in the future. But most of all, it finally makes us a public company. What that means is highly significant: PAETEC shareholders will now have shares valued at several times the price they would have fetched had we gone public at any time in the past. Those who have been waiting for years to exercise their options or cash in their shares can now enjoy the reward of all the work we've put into this company since we began.

The merger itself, though, is the perfect match. It isn't just an expansion into new economic territory for us. It's a homecoming of sorts. I worked with the co-founders of US LEC, Richard Aab and Tansukh Ganatra, long before PAETEC was founded. In fact, one snowbound day in 1996 at the Fairfield Inn near the Rochester airport, they invited me to join them in their plan to create US LEC. But I'd been traveling so intensely during several years with ACC, and for family reasons I wasn't able to move the clan to North Carolina. I couldn't justify the commute, and the negative impact on my family, so I declined. Now, a decade later, we've reunited, and it's fortunate that things have turned out this way. Building two separate companies, I think we've grown much faster than we could have grown if we'd teamed up at the start. It's yet another way of keeping it personal, because Tan trained many of our key people at PAETEC and most of them have great affection for both Rick and Tan.

Combined, I predict PAETEC will be the one company everyone turns to for telecom and data services as an alternative to dealing with one of the giants.

Yet, though our people will now finally be getting some of the rewards they have waited so long to enjoy, in many ways, the success of our transformation from a small startup into a larger, sustainable business model fit for a new technological world will be measured by how well some things *don't* change. How well, in fact, we stick to our four brand elements. The changes we go through may enable us to make business even more personal than ever before. We'll keep answering the phone with a live human voice when you want us to, but no matter who you speak with at PAETEC, soon that person *will* be able to know everything PAETEC knows about you and your situation. And no matter who you are, or how small your account, you'll still be able to reach me at arunas@paetec.com. It's all about staying in close touch with who you are and what you need. Whether or not we hear more from you, you'll be hearing more from us. Even after eight years, it feels as if we're finally just getting started.

About the Authors

ARUNAS A. CHESONIS

Mr. Chesonis serves as Chairman of the Board and Chief Executive Officer of PAETEC Corp., and is responsible for the vision, leadership, and direction of the company. Mr. Chesonis began his career at Rochester Telephone Corporation, now part of Citizens Communications Company. He went on to serve as President of ACC Corp., the parent company for all ACC-owned operations in the United States, Canada, Germany, and the United Kingdom, from February 1994 until April 1998, and was elected to its Board of Directors in October 1994. He holds a BS in Civil Engineering from Massachusetts Institute of Technology, an MBA from the William E. Simon Graduate School of Business at the University of Rochester, and an Honorary Doctorate of Laws from the University of Rochester.

DAVID DORSEY

David Dorsey is the author of *The Force,* which was selected as one of the ten best business books of the year by *BusinessWeek.* He is also author of *The Cost of Living,* a novel. He has contributed to magazines such as *Fast Company, Worth, Inc.,* and *Esquire.*

The Force is a literary non-fiction account of a year in the life of a top salesman and his team at Xerox Corporation. Published by Random House in 1994, and in paperback by Ballantine in 1995, it was hailed by reviewers as "an uncompromising portrait of a modern salesman." *The New York Times* selected it as an Editor's Choice. *The Cost of Living* was published by Viking/Penguin in 1997. Mr. Dorsey was a ghostwriter for Peter Georgescu on *The Source of Success,* published by Jossey-Bass/Wiley in August 2005. This is a book about how integrity and values have become central to business success in an excess-demand global economy.

Acknowledgments

THIS BOOK NEVER WOULD HAVE BEEN WRITTEN without the suggestion of Jeff Burke. He also brought us together as co-authors. As a result, we've both found this partnership fruitful and enjoyable. It has become a lasting friendship and working relationship and will lead us into new projects in the future.

We have to give a shout-out to Al Simone, who, as president of Rochester Institute of Technology, supported us without hesitation when we approached him. It's hard to imagine a better working relationship than the one we've had with RIT.

Many others have been creative and helpful, personally, with this book. Our PAETEC graphics team led by Steve Moose and several interns, including Elizabeth Schirmer and Sarah Wood, who kept this book moving along when other matters drew us away from it. Also the entire marketing team has to be thanked for the warmth they showed when the needs of research required their help and patience. Jack Baron, Tracy Robertson, and their team were champions of this book from the start and went out of their way to help arrange interviews, set up meetings, and alert us to events and people who deserved attention in the book. They also made the whole process fun.

Most importantly, we need to thank the dozens of people, at all levels of the company, who took time from their busy schedule to let us talk with them, follow them, watch them work with customers and fellow employees, and do it almost invariably with a smile. It isn't always easy having somebody look over your shoulder while you do your job, but you would never have known it from the warmth these people showed us. It's hard to imagine a nicer group of people to write about.

Finally, none of us can ever be successful without the patience of a partner at home. Arunas would like to thank his wife, Pam. The two of them have been together for seventeen years, and it has been her counsel that has kept him focused and grounded since they met. The company would never have been formed without her faith in him, and he will always be grateful to her. Dave would like to thank his wife, Nancy, whose enormous patience and love have seen him through some of the unique trials of working on this book. He'd like to thank, as well, his parents and children, who are invariably supportive and understanding when faced with the peculiar challenges familiar to those who have a writer in the family.

Index

Colophon

Typeset in Adobe Garamond Premier Pro
and Univers Condensed.

Printed on Nature's Natural 50%
post-consumer recycled paper.

Printed by Thomson-Shore, a member
of the Green Press Initiative.

First edition: 1,000 special edition
hardcover copies, 7,000 paperback.